PROPOSAL WRITING
The Art of Friendly and Winning Persuasion

William S. Pfeiffer
Southern Polytechnic State University

Charles H. Keller, Jr.
Keller Proposal Development & Training

Prentice Hall
Upper Saddle River, New Jersey Columbus, Ohio

Library of Congress Cataloging-in-Publication Data
Pfeiffer, William S.
Proposal writing : the art of friendly and winning persuasion /
William S. Pfeiffer, Charles H. Keller. Jr.
p. cm.
Includes bibliographical references (p.) and index.
ISBN 0-13-658213-3
1. English language—Rhetoric. 2. English language—Business
English. 3. Proposal writing in business. 4. Persuasion (Rhetoric)
I. Keller Jr., Charles H. (Charles Harold), 1949- . II. Title.
PE1479.B87P47 2000
808'.06665—dc21 99-33817
CIP

Editor: Stephen Helba
Production Editor: Louise N. Sette
Production Supervision: Carlisle Publishers Services
Design Coordinator: Robin G. Chuker
Text Designer: Laurie Janssen of Carlisle Communications
Cover Designer: Michael Osadciw
Cover art/photo: © Michael Osadciw
Production Manager: Matt Ottenweller
Marketing Manager: Chris Bracken

This book was set in Meridien by Carlisle Communications, Ltd., and was printed and bound by R.R. Donnelley & Sons Company. The cover was printed by Phoenix Color Corp.

Printed in the United States of America

10 9 8 7 6 5 4 3 2 1

ISBN: 0-13-658213-3

Prentice-Hall International (UK) Limited, *London*
Prentice-Hall of Australia Pty. Limited, *Sydney*
Prentice-Hall of Canada, Inc., *Toronto*
Prentice-Hall Hispanoamericana, S. A., *Mexico*
Prentice-Hall of India Private Limited, *New Delhi*
Prentice-Hall of Japan, Inc., *Tokyo*
Prentice-Hall (Singapore) Pte. Ltd., *Singapore*
Editora Prentice-Hall do Brasil, Ltda., *Rio de Janeiro*

Dedication

Sandy Pfeiffer thanks Evelyn, Zachary, and Katie for their support during the completion of this writing project.

Chuck Keller remembers Rodney Dabbs, a friend, mentor, and true proposal professional, and Chuck and Mercedes Keller, who were loving and supportive parents.

Introduction

The title, *Proposal Writing: The art of friendly and winning persuasion,* is notable for what it says—and doesn't say—about the contents of this book. This book is certainly concerned with proposal writing, but the act of writing is only one of many tasks required to develop a successful proposal. If you don't complete all the steps, a well-written proposal can fail.

This book was written as a textbook to guide students through all stages of the proposal process. It also is intended to provide working professionals with a reference book for writing proposals on the job. In either case, the book's goal is to give you the tools to write a successful—that is, a winning—proposal.

FEATURES OF THIS BOOK

The following features make this book a valuable tool for its dual audience:

- Although focused on proposals that seek commercial or government contracts, guidance is provided for sales letters, oral presentations, and not-for-profit proposals.
- Proposal development is described as an orderly and repeatable process. The sequence of the chapters supports a thorough understanding of this process.
- A systematic approach is described for managing the process—with emphasis on proposal management plans.
- Each chapter has a list of objectives and a chapter summary and presents proposal guidelines in a checklist-type format. These features reinforce key points and help make the book an effective teaching and reference tool.
- Numerous examples, marginal comments, and exercises amplify the book's basic instruction.
- Advice from many proposal experts is prominently displayed as cited quotations throughout the chapters. A bibliography identifies the source of these quotations and other material used in the book.

With so many types of proposals for so many types of customers, no one book—including this one—addresses all proposal issues. However, this book provides enough information to give readers all the basic skills to develop any type of proposal.

Let's now take a more detailed look at the structure and content of this book.

Chapter Summaries

Following are the main points covered in each of the book's 11 chapters.

- **Chapter 1, Marketing and Proposal Strategy**—conducting the proposal process as part of marketing, with emphasis on developing a proposal strategy and making the decision to pursue a proposal opportunity
- **Chapter 2, Proposal Management**—managing a proposal project, with emphasis on planning and following a repeatable process
- **Chapter 3, Preparing to Write**—planning the proposal with outlines, storyboards, and mock-ups before the writing begins
- **Chapter 4, Writing**—understanding your readers' needs and then meeting those needs with writing that is readable, accurate, responsive, and persuasive
- **Chapter 5, Graphics**—understanding the benefits of using proposal graphics and developing various types of graphics to enhance your proposal
- **Chapter 6, Sales Letters and the Executive Summary**—writing sales letters that support your proposal and producing what can be one of the most important parts of your proposal—the executive summary
- **Chapter 7, Formal Proposals**—developing longer and more complex proposals, known as formal proposals
- **Chapter 8, Informal Proposals**—developing shorter and less complex proposals, known as informal proposals
- **Chapter 9, Reviews**—conducting effective internal proposal reviews, with emphasis on following a review plan and offering constructive criticism
- **Chapter 10, Editing**—completing the three stages of proposal editing—making stylistic changes, correcting grammatical errors, and proofing to correct mechanical errors
- **Chapter 11, After Proposal Submittal**—executing duties that occur after the written proposal is delivered, including making oral presentations, responding to customer questions, negotiating, and performing various post-award tasks

To better understand this book, readers should be familiar with key proposal trends and definitions. First, let's examine current trends in proposal writing.

Proposal Trends

Proposal development has undergone major changes over the last two decades. These changes have affected the resources and methods used to write proposals.

Proposal Development as a Profession. Those who develop proposals are being recognized as skilled professionals who add to the profit of their companies. Their work is increasingly seen as a specialized discipline for full-time positions, not just a collateral job. Dedicated proposal organizations are now common in businesses. The demand has grown for proposal consultants and contractors to supplement the internal proposal resources of businesses. Reflecting the growing status of the discipline, the Association of Proposal Management Professionals (APMP) was formed in 1989 and has a growing membership list of proposal professionals in the United States and abroad.

International Competition. American businesses are competing more than ever before in the international market. In addition to producing proposals for domestic—and English-speaking—customers, they are submitting sales proposals to foreign customers. This international market presents businesses with many cultural challenges, from translating the proposal into a foreign language to printing it on paper sized differently than it is for the domestic customer.

Advances in Computer Technology. New, more powerful—and less expensive—hardware and software provide better tools for managing, writing, illustrating, and publishing proposals. Personal and laptop computers have become standard tools for the proposal team. Yet it wasn't long ago that a writer produced a proposal draft in longhand and then passed it to a word processing pool for typing.

Advances in Communication. Communication has improved with the advent of Internet, intranet, and telecommunication technology, making it easier for proposal team members to communicate among themselves and with their customers. In particular, the Internet provides a forum for selling products and services, distributing solicitations for proposals, and submitting bids. The use of "virtual" proposal development centers, accessible through Internet or intranet web sites, can allow on-line coordination from one location even if team writers are scattered throughout the world. Electronic mail (e-mail), facsimile (fax), and overnight shipping also allow proposal teams to exchange information quickly.

Proposal Definitions

Let's now define some important terms you'll see in this book, as well as in the real world of proposal development. We'll do so with answers to five basic questions about proposal development.

Question #1: Has the proposal been requested?

- **Unsolicited proposal**—If you submit a proposal that's not requested by its recipient, the proposal is unsolicited.
- **Solicited proposal**—If the proposal is requested by its recipient, the proposal is solicited. The request may come informally by phone or in face-to-face conversation, or it may come through a written solicitation—a request for proposal (RFP).

Question #2: Is the proposal to a prospective customer or to your boss?

- **External proposal**—If the proposal is directed to someone outside your organization, it's an external proposal.

- **In-house proposal**—If the proposal is directed to someone within your organization, it's an in-house, or internal, proposal.

Question #3: What are you "selling" in your proposal?

- **Sales proposal**—If the external proposal attempts to sell something, it's a sales proposal—the most important type of external proposal used in the business world.
- **Grant or R&D proposal**—If the external proposal seeks funding to perform research or to develop a product, it's a grant or research and development (R&D) proposal. An R&D proposal can be internal if it requests funding from your own organization.
- **Planning proposal**—If the proposal requests support for a change of internal procedures, it's a planning proposal.

Question #4: What proposal format is most appropriate?

- **Formal proposal**—Whether for an internal or external reader, if the proposal is more than five pages, excluding attachments or enclosures, and adopts a more formal format, it's a formal proposal. It can be written in multiple volumes and include thousands of pages. A formal proposal is expected to have multiple sections, detailed content, a table of contents, binding, and a cover.
- **Informal proposal**—If the proposal is short and relatively uncomplicated—no more than five pages excluding attachments or enclosures—it's an informal proposal. Unlike the formal proposal, it isn't expected to have a binding or a table of contents. If the informal proposal is for an internal reader and follows a memorandum format, it's a memo proposal. If it's written for an external reader and follows a letter format, it's a letter proposal.

Question #5: Is your proposal correspondence external or internal?

- **Sales letter**—Proposal-related correspondence to someone in an external organization is called a sales letter. Because these letters can strengthen relationships—before and after proposal submittal—they are used to increase the likelihood of getting work both now and later.
- **Sales memo**—Proposal-related correspondence to someone within your organization is called a sales memo. Although you often communicate in person with an internal audience, a sales memo can be used to support an internal proposal before and after proposal submittal.

Your company or prospective customers may use other terms or definitions, so be ready to adapt. Regardless of what you call your proposals or who is in your proposal audience, your proposal should have one basic goal: to convince the proposal readers that you can meet their needs. This book will help you attain that goal.

Proposals are either winners or losers; we hope all of your proposals are winners.

ACKNOWLEDGMENTS

We would like to thank the many people who have provided us with the knowledge that helped us write this book.

Sandy Pfeiffer expresses his appreciation to Fugro-McClelland, Inc., in Houston, Texas. He also wishes to thank the following friends and colleagues for material and suggestions used in this book: J. Lundy, Betty Oliver Seabolt, Hattie Schumaker, Herb Smith, James Stephens, and Tom Wiseman. Chuck Keller thanks his colleagues in the APMP and the many professionals he has worked with at the Lockheed Martin Aeronautical Systems in Marietta, Georgia.

In addition, we thank Verna Hankins, Rob Kelly, and Tom Stover for their assistance and the following students or graduates for permission to use their work as the basis for examples: Sam Rundell, David Cox, and Rob Duggan.

We also thank Prentice Hall for allowing us to use examples and material from other textbooks written by Sandy Pfeiffer:

- *Technical Writing: A Practical Approach (3rd edition, 1997, and 4th edition, 2000)*
- *Proposal Writing: The Art of Friendly Persuasion (1989)*

Contents

PROPOSAL WRITING
The Art of Friendly and Winning Persuasion

1 Marketing and Proposal Strategy

OBJECTIVES

- **Understand the importance of building strong personal relationships with those to whom you write proposals**
- **Learn to cement the bond between you and the readers of your external sales and in-house proposals**
- **Learn to reach an informed decision on whether you should submit a proposal**
- **Learn to develop a winning proposal strategy**
- **Perform exercises using chapter guidelines**

On a sunny Saturday afternoon, Max Smith is washing his car in his driveway. Halfway through the job, his next-door neighbor's daughter, an assertive 10-year-old, walks over to the edge of the drive. Susie has magazine brochures in one hand and a pencil in the other, so Max quickly surmises that it's time once again for the school's annual fund-raising drive. Although he already receives more magazines than he can possibly read, Max finds himself cheerfully ordering two more subscriptions. After all, Susie needs just 10 more sales to earn a portable CD player for top salesperson of the week!

Let's analyze this sale. Why is Max such a pushover? He regularly turns down phone solicitations for magazines and tears up magazine sweepstakes mail without even opening it. The answer, of course, is that he knows Susie and her parents. He'll buy a product from his neighbors that he wouldn't consider buying from someone he doesn't know.

This book starts with the simple principle evident in Max Smith's purchase. Your proposals are most successful when you know your audience. Getting to know your proposal readers means discovering and then satisfying their deep-felt needs. This goal is the stuff of long-term trust building. Indeed, it supplies the driving force behind all proposal writing.

This book assumes that your proposals aim to satisfy the needs of (1) customers (also known as clients) outside your firm, to whom you propose a product or service, or (2) managers in your own organization, to whom you propose in-house changes. In either case, you need to be interested in the long-term goal of building good relationships and securing repeat business from a satisfied customer.

BUILDING PERSONAL RELATIONSHIPS

Whatever you're selling, remember that your goal is to strengthen personal relationships. This process relies on the strategy of establishing a pattern of repeated and helpful contacts with the people you are trying to influence. Some contacts are quite personal, as in meetings or phone conversations, while others, such as letters, are less so. All of them work together to convince your customers or your managers that you can satisfy a need.

View the proposal process as a continuum of contacts with your audience. This process begins with your first meeting, phone call, or letter and ends with the last thank-you letter or project status report. The "end" of the relationship is only temporary, of course, because successful projects and strong working relationships can lead to more work.

In this chapter, we will outline two communication cycles. One applies to your customers and thus to sales proposals; the other applies to your managers and thus to in-house proposals. The key to your proposal success is your persistence in following through on all these steps. Then we'll take a close look at two processes that support these cycles: the decision to develop and submit a proposal and the analysis and planning needed to develop winning proposal strategies.

THE SALES PROPOSAL PROCESS: 12 STEPS TO SUCCESS

> No two better allies exist in the new business arena than marketing teams and proposal teams. These two teams are involved in a unique, symbiotic relationship, in which both prosper or suffer according to the acceptability of a stack of pages called the *technical proposal*. . . . However, despite this vital interdependence, marketing and proposal groups are separate entities in many companies.
>
> (Whalen, 1996, p. 1-1)

The success of your marketing and sales efforts depends on giving your customers the personal attention they deserve. Excellent service means being there, ready to help them analyze and then solve their problems. As a reward, you establish a strong bond with them. It's this long-term relationship that companies should seek to establish. Your proposals have a much better chance when directed to someone who already respects your work, rather than to a new customer.

The sales proposal process isn't just the responsibility of the proposal team. It's the duty of anybody in your organization whose job is to identify and meet customer needs. The marketing, sales, and proposal organizations should coordinate their efforts to form a strong relationship with customers as early as possible in the sales proposal process.

There are 12 steps that will help cement the bond with your customer and generate repeat business. Figure 1–1 summarizes this process. The steps, key elements of effective marketing and sales plans, are grouped into four stages in the sales proposal process:

- Stage #1—Breaking in
- Stage #2—Getting known
- Stage #3—Clinching the job
- Stage #4—Following through

> Marketing research is a crucial ingredient of the marketing plan. Every corporation should know the answers to these five questions:
>
> 1. Who is the target audience?
> 2. What do the customers want?
> 3. What does the competition offer them?
> 4. What can we offer them?
> 5. What do they think we offer them?
>
> (Luther, 1992, p. 131)

Although the most likely context for this process is a solicited (requested by the customer) sales proposal, the steps can be tailored to accommodate a proposal that is unsolicited (not requested by the customer). The implementation of the 12 steps can also be affected by the type of customer. For example, the U.S. government customers can be influenced by laws, regulations, and agency procedures that have no application to commercial customers.

Stage #1, Breaking In

■ *Step #1: Send a letter to a prospective customer.*

Begin the marketing process with a letter to the potential customer. Even if it's a "cold-call letter" to someone you found on a mailing list, a first letter is a necessary start if you've had no previous contact with the customer. Chapter 6 offers suggestions for writing letters, including "cold-call" letters. For now, remember not to seek any commitments in this first letter. You're dealing with people who need time to evaluate your product or service. For example, avoid asking readers to return a card or form, a request that has a notoriously low return rate. Instead, state in the letter that you will give them a phone call.

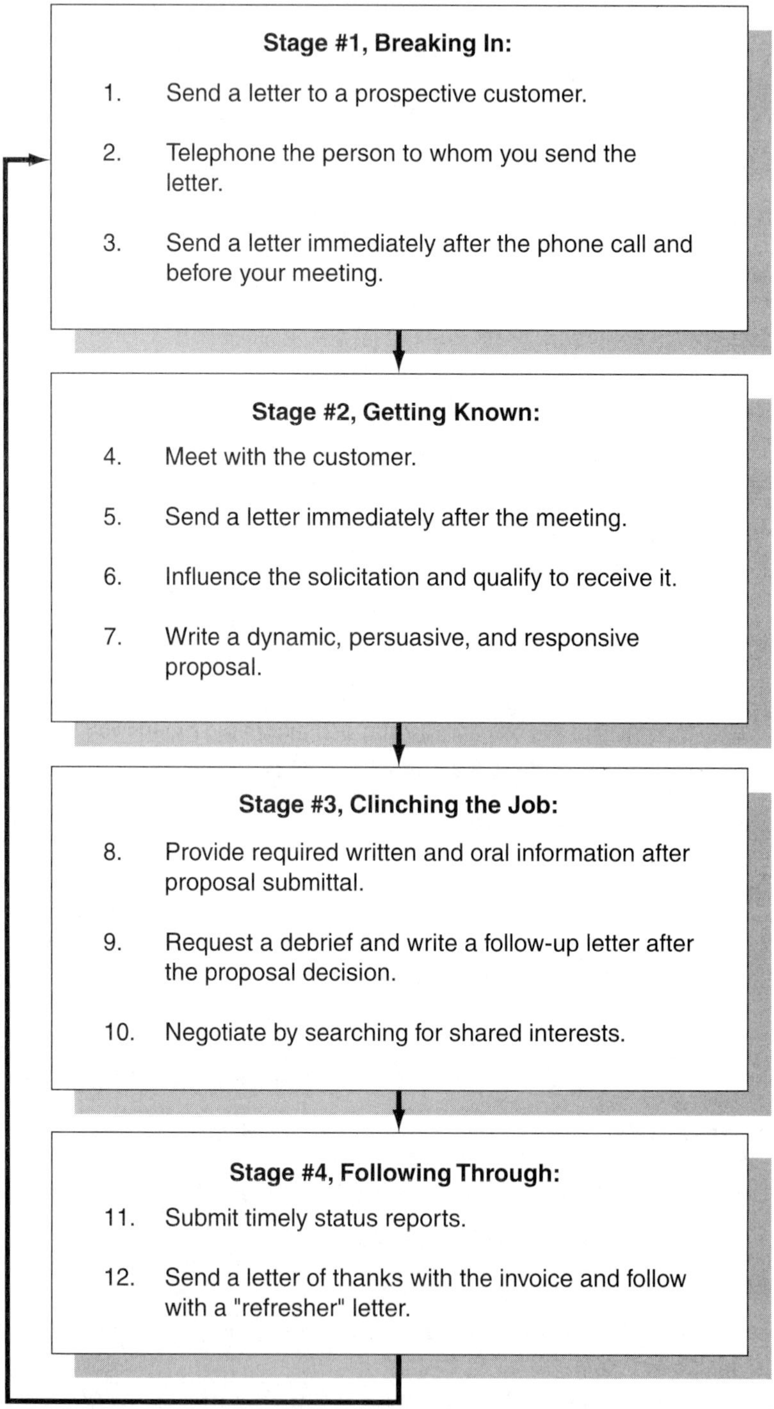

Stage #1, Breaking In:

1. Send a letter to a prospective customer.
2. Telephone the person to whom you send the letter.
3. Send a letter immediately after the phone call and before your meeting.

Stage #2, Getting Known:

4. Meet with the customer.
5. Send a letter immediately after the meeting.
6. Influence the solicitation and qualify to receive it.
7. Write a dynamic, persuasive, and responsive proposal.

Stage #3, Clinching the Job:

8. Provide required written and oral information after proposal submittal.
9. Request a debrief and write a follow-up letter after the proposal decision.
10. Negotiate by searching for shared interests.

Stage #4, Following Through:

11. Submit timely status reports.
12. Send a letter of thanks with the invoice and follow with a "refresher" letter.

FIGURE 1–1
Sales proposal process

■ *Step #2: Telephone the person to whom you send the letter.*

Telephoning dramatically increases your chance of finding good leads. You become a real person to the customer, not merely a purveyor of junk mail. More important, a phone conversation gives you the first opportunity to demonstrate how you can satisfy a need. Show yourself to be someone who is trying to help solve a problem, not just trying to sell something. In our example, the caller begins with a point that will probably catch the listener's attention: "Mr. Sampson, I mentioned in my letter that we've found a number of toxic waste sites on commercial property in Barnes County. I thought you might want to know about some of these sites, in case your bank plans to invest in that area."

The tone here is helpful, not pushy. The goal is to refresh Sampson's memory about your letter, get his interest in a topic that might concern his bank, and offer useful (and free) information. Later in the relationship, you can become specific about your firm's interest in conducting toxic waste surveys on commercial property for which his bank might provide mortgages. For the moment, however, you need to help him realize there's a problem that needs to be addressed. If the call goes well, you'll arrange a time to call back after he has thought about what you're offering. If the call goes really well, you'll want to arrange a meeting with Sampson.

■ *Step #3: Send a letter immediately after the phone call and before your meeting.*

Your letter may include some or all of the following objectives:

- Express some excitement about the ideas the two of you discussed on the phone.
- Summarize the results of the discussion.
- Offer a few more ideas you've put together since the call.
- Confirm the arrangements for your meeting.

Again, the goal is to be seen as the "helper." You want to reinforce whatever good communication may have begun between the two of you over the phone. This letter must be in the mail right away—no later than 48 hours after the phone call. The real value of all sales letters is their immediacy.

Stage #2, Getting Known

■ *Step #4: Meet with the customer.*

This book is mainly about written and oral presentations, not interpersonal sales techniques. It's worth noting, however, that your first meeting with the customer must not be a stressful or forced affair. Be a professional who has answers to problems, not a slick salesperson with a product or service to pitch. Maintain a tone and approach that demonstrate your interest in diagnosing the customer's problem, without prematurely discussing what you have to sell. The following questions reflect the approach and tone you might want to adopt:

- "Mr. Sampson, are you familiar with your bank's liability if you were to foreclose on commercial property with toxic waste problems?"

- "Could you give me a few more details about the kinds of commercial property First National is interested in?"
- "What kinds of information do you now get about the use and condition of commercial property for which you might provide loans?"

Notice that these questions encourage Sampson to talk about his problems. Listen carefully and take notes. When it seems appropriate, of course, you can interject comments about your company's product or service, but only when the need for them has been fully explored.

Arrange a meeting with a goal in mind. Your objective might be simply to introduce yourself or your company to the customer. Or if you're expecting a formal request for proposal (RFP) from the customer, you might ask for details about RFP requirements, ask the customer's opinion on your company's approach to those requirements, or give advice about what those requirements should be. The use and content of RFPs are described in Chapters 3 and 7.

Take notes during the meeting and then afterward write and file a trip report about your observations. The information can be used to develop strategies for your proposal, especially for those on the proposal team who have little direct contact with the customer, and to provide "between the line" information for interpreting the RFP. Do research about your customers; read associated industry publications, visit their Internet web sites, read old RFPs issued by them, and review proposals your company has already submitted to them.

When seeking U.S. government work, pre-RFP meetings can also provide you with the best opportunities to talk with government representatives. After RFP release, contact with them may be limited and tightly controlled.

Successful first meetings can end in two ways. You may be fortunate enough to be asked to submit a proposal, or you may leave with the understanding that you are to provide the customer with more information. This approach helps you maintain control, even though the customer is still undecided about your product or service.

■ *Step #5: Send a letter immediately after the meeting.*

After the meeting, send a follow-up letter to summarize briefly your understanding of the conversation and to give you the opportunity to send along a helpful item—maybe a marketing brochure or product information sheet. Don't squander these items by leaving them in the customer's office when you visit. Save them for the follow-up letter, because they may get more attention and give you good reason to make this additional contact with the customer.

At this point, you may be wondering if you can overdo this business of writing letters. But remember that it's human nature to look forward to receiving first-class mail, or even electronic mail (e-mail). A letter remains one of the cheapest, most effective ways to reach the decision makers in any organization.

Having written your letter, you may engage in another series of phone calls and meetings before you receive the request for an official proposal. This request can come in a formal RFP, or simply from a phone call asking you for cost figures and rationale. Later in this chapter, we describe how to decide if you should respond with a proposal.

■ *Step #6: Influence the solicitation and qualify to receive it.*

During Stage #2 (Getting Known), customers may also ask you to provide information that can help them better define requirements in their RFP or to determine if your company is qualified to bid on the RFP. These pre-RFP inputs can come from the following activities:

- Pre-RFP meeting—The customer can schedule a meeting for contractors to discuss the upcoming RFP or contract opportunity, and to solicit comments about a draft RFP or a yet-to-be developed RFP. The meeting can help you learn more about the contract requirements and influence RFP requirements. If there are other contractors in attendance, you might identify your potential competition and make contacts with other contractors for possible teaming agreements.
- Request for information—The customer can summarize its needs and ask contractors to submit written ideas on how these needs might be met. The request might be structured like an RFP and the contractor response might have the appearance of a proposal. This response can help the customer better define its requirements and prepare contractors for the release of a formal RFP. For the contractor, it's a chance to influence contract requirements and to begin pre-RFP proposal planning.
- Request for draft RFP comments—The customer can release a draft RFP to contractors for their written comments about the document. These comments can include clarification questions and recommendations for RFP changes, additions, and deletions. This step can help both the customer and contractor: the customer can better define its requirements, and the contractor can influence those requirements and begin pre-RFP proposal development. You can get an early start on your proposal by using the draft RFP to develop a proposal draft before the formal RFP is released.
- Request for a qualification statement—The customer can ask contractors to submit a written summary of their abilities, products, services, and past experience. This summary can be used to judge your ability to meet the contract requirements and to determine if you'll receive the RFP. This qualification statement might be requested for a specific RFP or be periodically placed on file with the customer to automatically place you on the customer's RFP distribution list. In contrast, you might simply need to ask for an RFP without any details about your qualifications. Check with your customers to learn what criteria they have for placing contractors on their RFP distribution list.

■ *Step #7: Write a dynamic, persuasive, and responsive proposal.*

With the release of an RFP, the extent of customer-contractor contact can be closely limited by the customer. This approach allows the customer to maintain administrative control of the procurement process in a manner fair to all competing contractors. In the commercial market, there may be little or no limit to this contact. However, when pursuing government work, especially with the U.S. government, expect strict controls dictated by laws and regulations. Comply with all contact limits spelled out in the RFP and check with the customer to resolve any questions you have about the limits.

Customers often give contractors the option of submitting written questions about the RFP. Expect a submission deadline, which allows time for the customer to produce and distribute the answers and for the contractors to reflect the answers in their proposals. Responses to these questions can be in the form of written and/or oral answers. After RFP release, the customer may conduct a bidders' conference to provide answers to the written questions received prior to the meeting and to oral questions raised at the meeting. You can also expect—especially with the U.S. government customer—that all written questions and answers will be distributed to all contractors, regardless of the question source.

After RFP release, the customer may issue written amendments to change, clarify, or add RFP requirements. An amendment can result from many factors, including customer responses to contractor questions, changes in customer requirements, mistakes in the original RFP, and the customer's decision to extend the deadline for proposal submittal.

Proposal development is normally a collaborative process that demands planning, strategy, and sound proposal process to produce a proposal that meets the customer's needs. Later in this chapter, we will describe how to develop a winning strategy for your proposal. Chapter 2 describes the management of the proposal process, while Chapter 3 explains the prewriting steps to develop the proposal outline, storyboards, and mock-ups. The writing, art development, and editing of the proposal are described in Chapters 4, 5, and 10, respectively. Chapter 4 gives particular attention to analyzing the audience and writing to meet its needs. Chapter 9 stresses the need to respond to customer requirements by conducting internal proposal reviews.

The success of your written proposal also demands the relationship-building efforts that support it—letters, meetings, and phone calls. In the last analysis, your proposal stands as a statement about you as a communicator.

Stage #3, Clinching the Job

■ *Step #8: Provide required written and oral information after proposal submittal.*

After submission, be prepared to provide additional information to the customer. This information can be crucial in winning the contract, so carefully plan, prepare, and respond within the allotted time. Chapter 11 provides more details about these postsubmittal activities.

During their evaluation of the written proposal, customers may ask you to clarify a statement in your proposal or to provide information about a topic that wasn't adequately covered. The customer may also ask you to submit a written best and final offer (BAFO), which gives you the chance to change your cost proposal (the general implication: lower your price).

Customers may also require an oral presentation for you to summarize your written proposal and to allow them to ask you face-to-face questions. The customer may require you as a bidder to demonstrate your product or to play host to a tour or inspection of your facility by its representatives.

■ *Step #9: Request a debrief and write a follow-up letter after the proposal decision.*

After a contract award decision is made—whether you win or lose—ask customers for a debrief of the strengths and weaknesses of your proposal and send them a follow-up letter.

If your proposal was rejected, a letter thanking the customer for the opportunity to bid can show that you're a good "sport" and that you're interested in keeping the door open for later work. You may also want to ask why your proposal was rejected or how it was outmatched by your competition. The desire to know your own deficiencies can indicate that you're the kind of flexible, concerned person or firm the customer should consider for the next contract. Remember that you're in the business of building long-term relationships, not just getting one contract.

Supplement the customer debrief or other feedback with a self-assessment of your proposal strengths and weaknesses. Collect all postdecision analyses into a lessons-learned file, which can be used for future proposals to the same or different customer.

In the case of acceptance, express appreciation for the award decision. Write your letter to reinforce the idea that the customer has made the right decision.

■ *Step #10: Negotiate by searching for shared interests.*

Many sales arrangements require negotiations to set final terms of the contract. As explained in Chapter 11, this is the time to abandon old ideas about negotiating, which has progressed beyond the intimidating "I win/you lose" techniques of the past. Instead, be consistent with the tone set thus far in the proposal process by finding out what you have in common with the customer. After establishing points of agreement, you're in a better psychological position to pursue areas about which there may be differences of opinion. This gentler approach to negotiating doesn't mean giving in too quickly or failing to assert yourself. Instead, it means striving to have customers view you as a colleague rather than an adversary.

Stage #4, Following Through

■ *Step #11: Submit timely status reports.*

As required by the negotiated contract, provide all contract/project status reports to meet schedule, costs, technical, management, and quality requirements. Strictly speaking, reports to the customer are not part of the proposal process and thus are not covered in this book. But never forget that all business writing, including reports, should be persuasive. For example, you can convincingly show your capability with reports showing the project is progressing on schedule or that it was completed on time after meeting all customer expectations.

■ *Step #12: Send a letter of thanks with the invoice and follow with a "refresher" letter.*

Billing can be authorized incrementally during the contract or after the contract is completed. Unfortunately, the last impression your customers might have of you could be your request for money. If appropriate for your contract and customer, soften the impact of your billing request by sending a brief letter of thanks with the invoice. In the letter, show your interest in future work. If the invoice goes to an accounting department rather than to your technical or management contacts, also send them a separate letter of thanks. Leaving customers with good feelings about your efforts can pave the way for your being selected again.

To keep good customers, you need to restart the sales proposal process after the invoice letter. To remind customers of your company, send them a "refresher" letter no more than six months later or whenever you learn about a new contract opportunity with those customers. In this letter you should (1) say that you enjoyed working on the previous job and (2) set the stage for a new round of meetings, phone calls, letters, and a new proposal.

You can achieve the best proposal-acceptance ratios with customers you've served well in the past. Most marketing energy, therefore, should go toward getting repeat business. Then the sales proposal process is repeated.

THE IN-HOUSE PROPOSAL PROCESS: FIVE STEPS TO SUCCESS

We've explained that sales proposals result from careful attention to an entire marketing cycle. The same technique works for in-house proposals. To be sure, routine suggestions are often approved or rejected without the fanfare of a proposal or multiple contacts with managers. However, once your idea moves beyond the routine, make your best case by putting it in writing and plotting a complete internal marketing strategy. Use the following five steps to achieve success with your in-house proposal. Figure 1–2 shows the flow of this process.

■ *Step #1: Meet with your boss.*

After giving careful consideration to your idea, approach your supervisor in person to get an initial response. You may have some preliminary ideas to present in written form—for example, a one-page outline. If your supervisor is cool toward the idea, then you can back off without having made any major investment of time. If your supervisor shows some interest, however, then offer to write the proposal. Because in-house proposals often result from "inspiration" and are thus unsolicited, this one-on-one meeting can help you get your boss to think your way and set the stage to accept your written effort.

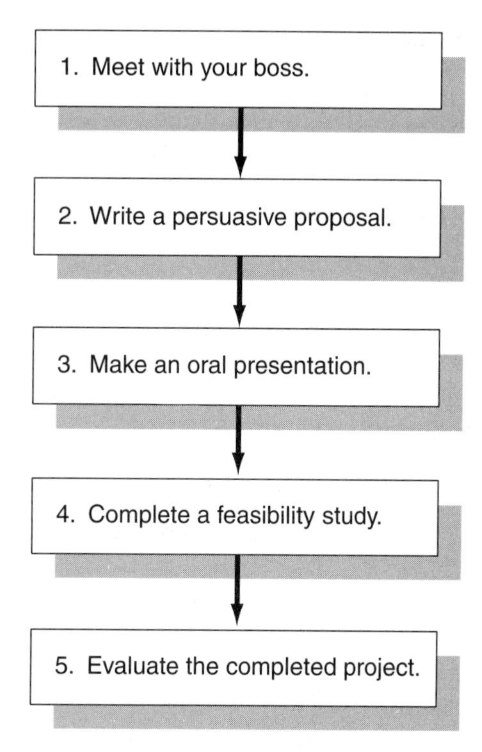

FIGURE 1–2
In-house proposal process

■ *Step #2: Write a persuasive proposal.*

The proposal should follow the in-house guidelines described in Chapter 8, unless your potential readers request another format. Before you submit the proposal, ask several of your trusted peers to review a draft of the proposal. You may discover that their differing views of the company's needs will lead to some useful advice. Just as important, these readers may acquire some "ownership" of the proposal. If they do and their opinions are later sought by those charged with accepting or rejecting your proposal, you will have already developed supporters. Also, solicit opinions of any employees who will be affected by your proposal, for these reasons:

- Like the peers whose views you seek, these colleagues can provide constructive criticism.
- They will be more likely to speak favorably about your idea later during the actual review process if they've had the benefit of an early reading.

■ *Step #3: Make an oral presentation.*

Offer to meet with decision makers to present highlights of the written proposal and to answer any questions about the proposal.

Step #4: Complete a feasibility study.

If the proposed project is approved but perhaps needs further research, offer to complete a feasibility study. Like a proposal and a report, a feasibility study evaluates the practicality of doing what you have proposed. Because the project will be associated with your name, give it a good chance of success before you proceed. An objective, thorough feasibility study should

- Establish clear, preferably measurable, criteria by which the proposed idea can be judged
- Include cost as well as personnel requirements
- Examine several alternatives besides the preferred one
- Support the alternative that best satisfies the established criteria, even if it's not the one you originally proposed

Step #5: Evaluate the completed project.

If the feasibility study proves positive and the project goes forward, be sure to evaluate the results. This evaluation should include a written report to those who approved the proposal, so that they are aware of the success of your work. Here are some questions that your evaluation should answer:

- What were the specific objectives of the proposed project?
- What objectives were reached?
- What objectives were not reached and why?
- Was the proposed schedule followed?
- Did the project stay within its proposed budget?
- Will follow-up work be needed and who will do it?

Even if the proposal resulted in a less-than-successful project, you can gain respect through your willingness to evaluate the project honestly.

THE BID/NO BID PROCESS: AN INFORMED DECISION

> Proposal writing is a most important part of marketing, but a great deal of work must precede the proposal writing. And the successful marketer will review a great many opportunities to write proposals, more than he or she can or should respond to. But that raises the question of the need to make the inevitable bid/no-bid decision, itself a most important element of marketing.
>
> (Holtz, 1986, p. 88)

Whether you are planning a sales or in-house proposal, make an informed decision to submit the proposal and develop winning proposal strategies once you've decided to submit the proposal. To demonstrate these steps, let's describe how they would be performed for a sales proposal in response to an RFP.

Writing proposals costs money, which is a good reason for you to reduce your risk of failure by building relationships with the customer before the RFP arrives (or before submitting an unsolicited proposal). Before rushing off to write the proposal, make an objective bid/no bid decision based on the evaluation of specific criteria. At a minimum, the decision should reflect the answers to the questions in Table 1–1.

Develop measurable standards for evaluating answers to the questions in Table 1–1 and require that a minimum grade or threshold be met before a bid

TABLE 1–1: Bid/No Bid Criteria and Questions

Criteria	Bid/No Bid Questions
Competition	• Do your qualifications for the proposed product/service match or exceed those of the likely competition? • Based on the expected selection criteria of the customer, how will your proposal be evaluated in comparison with your likely competition?
Contract	• Is it likely that your proposed contract provisions and costs for the product/service will be acceptable to the customer?
Future	• Could the proposal or a resultant contract lead to later contracts with the targeted customer or other customers?
Interest	• Is the project an interesting, challenging, and prestigious one for your company?
Goals	• Does the project fit within your company's long-range business plan?
Risk	• What level of risk do you see in your ability to meet the technical, management, schedule, and cost requirements for the proposed product/service?
Profit	• What is the expected revenue and profit for the contract, and does the profit potential meet your firm's needs?
Readiness	• How much bid and proposal funding is required to produce the proposal, and is the money available? • Do you have available proposal resources to submit a competitive proposal, and will you have project resources to deliver the required product/service if you win the contract?
Responsiveness	• Do you understand the customer requirements, and can you meet all the customer requirements for the product/service? • Are you prepared to commit the necessary resources to provide the required product/service on time and within budget?
Trust	• Have preproposal contacts helped build a trusting relationship with the customer? How much influence did you have in the making of the solicitation?
Uniqueness	• Can you offer unique products, services, or personal qualifications for the project?

decision is made. For example, there might be a requirement that the contract yield at least a 10 percent profit, or that the projected proposal cost not exceed the available bid and proposal (B&P) budget. For other questions, there might be a requirement to answer yes for certain questions or have a minimum percentage of yes answers for a group of questions. Other questions might be answered on a 1 to 10 scale with 1 being the least favorable and 10 most favorable, and require a minimum average answer before you decide to bid.

The format and scope of the questions and answers, the method and formality of the analysis, and the identification of decision makers in the bid/no bid process can take many forms. Regardless of the approach taken, the process should lead to a decision based on a sound business approach, not on intuition or "gut" feelings. You must be convinced that you can produce a competitive proposal for new business that fits your company goals and capabilities.

> It is a common mistake to believe that the number of contracts you win is directly proportional to the number of proposals you send out. Nothing could be less true. Rather, the number of contracts you win is a result of the professionalism and quality of your proposal planning and preparation. Excellent proposals represent investments of time and other resources. You therefore need to select your opportunities carefully.
>
> (Svoboda and Godfrey, 1989, p. 30)

THE STRATEGY PROCESS: SEVEN STEPS FOR A WINNING PLAN

When a decision to bid is made, focus on the development of strategies for writing and pricing a winning proposal. The strategies should come from an honest look at your ability—and your competition's ability—to provide the product or service. They should also be based on an analysis of key concerns, issues, and risks (technical, management, cost, or schedule) that are, or could be, your obstacles in providing the product or service.

> In the military, strategy is the grand plan by which a commander attempts to exploit the enemy's weaknesses by capitalizing on his own strengths. In other words, strategy is how you maximize your chance to win. It's the same basic thing in proposal writing. Strategy for you as a proposal writer is the intelligent application of your understanding of three things:
>
> 1. The client's business needs
> 2. Your business's strengths
> 3. Your competitor's strengths and weaknesses
>
> (Sant, 1992, p. 94)

Begin the strategy process by forming a strategy team. The composition of your strategy team can depend on many factors, including the company's organization, product or service to be proposed, and prospective customer. Avoid a team consisting just of high level executives who will have little or no direct role in developing the proposal. Ensure that the proposal team is well represented on the team, thus providing a link between strategy development and the proposal management and writers who must implement the strategy in the proposal. If possible before the first meeting, provide the strategy team with worksheets so they can begin listing ideas to discuss. You may find that several meetings will be needed to do a complete strategy analysis and plan.

Early strategy planning can provide the following benefits:

- If the bid/no bid decision is still open to discussion, a detailed and objective discussion of issues and strategies may lead to a firm decision.
- It can help you prepare for future contact with the customer, including pre- or post-RFP meetings, as well as comments and questions you may submit about a draft or formal RFP. With early strategy analysis, you lay the groundwork for better understanding customer needs, asking more effective questions, getting customer feedback about your proposed approach, and offering changes to the RFP that are to your advantage.
- Based on your strategies, you can refine baseline approaches for the management, technical, and cost strategies of your proposal. Simply stated, proposal baselines are "game plans" you will follow as you write your proposal.
- With your strategies and baselines in place, this early planning can save you time later in the proposal process—whether you decide to write a pre-RFP proposal draft or wait to begin a first draft after the formal RFP arrives.

Figure 1–3 summarizes the inputs to the strategy plan. To further describe the strategy plan process, let's assume your company reaches a bid decision to pursue a long-term contract to manufacture and deliver bicycles for a large sporting goods store. Before (Stage #1) or shortly after the RFP is released (Stage #2), company management decides to build upon the bid/no bid analysis by conducting a winning strategy meeting. At this meeting (or meetings, because it may take some time), perform the following steps.

Strategy Step #1: Identify your key obstacles and make strategies to overcome them.

List your key concerns, issues, and risks as you assess your ability to provide the requested bicycles, and then develop a strategy to mitigate or resolve each point.

EXAMPLE

- Your company's production line capacity can't produce the required number of bicycles during the first year of the contract.

  STRATEGY: Use manufacturing subcontractors to supplement your first-year production and, in advance of proposal submittal, make subcontract agreements with qualified subcontractors to provide the production support.

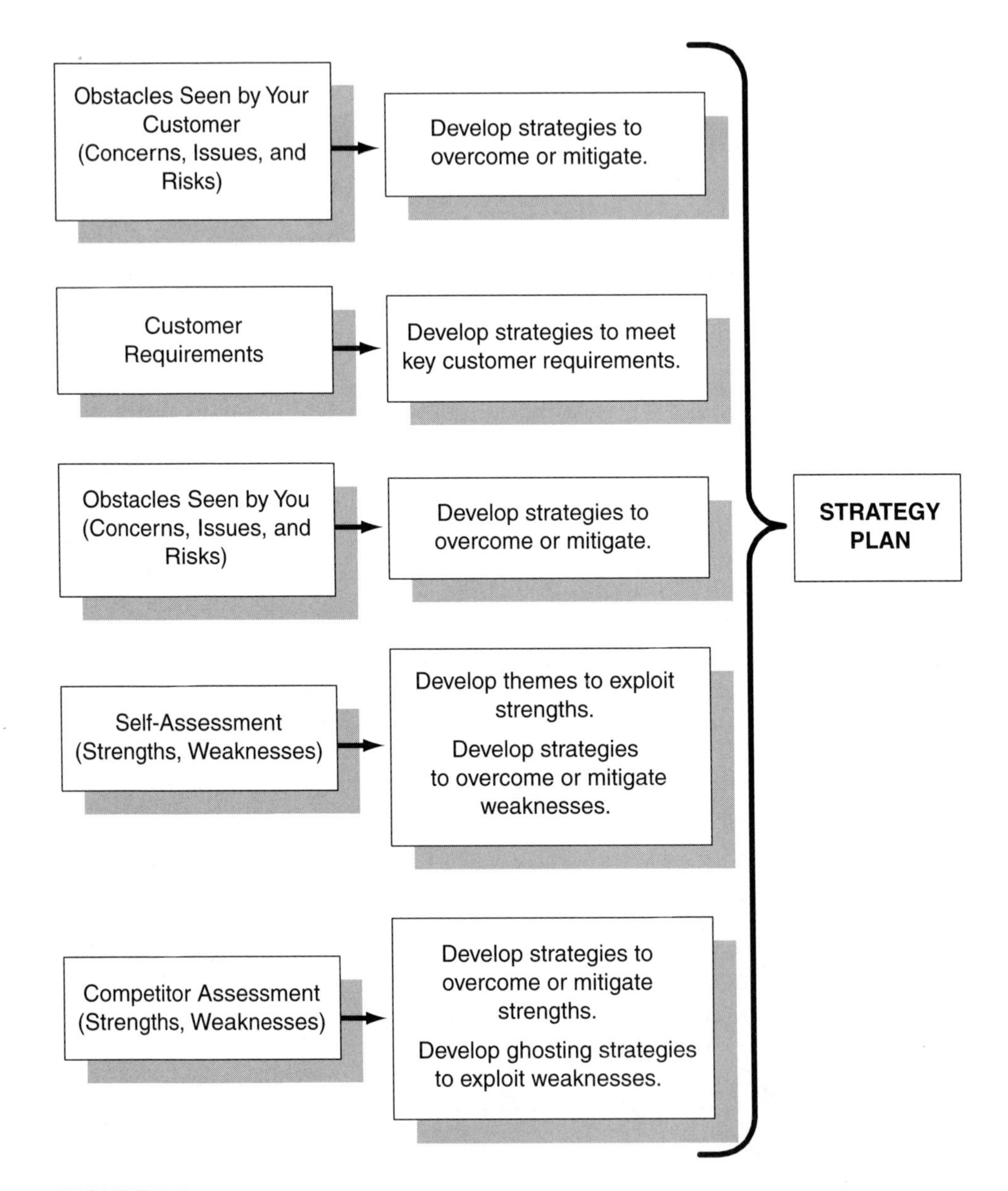

FIGURE 1–3
Inputs for the strategy plan

■ *Strategy Step #2: Identify key customer needs and make strategies to meet them.*

List the key concerns, issues, and requirements that affect how the customer will evaluate your proposal. Then develop a strategy to resolve or mitigate each concern, issue, or risk and to meet the customer requirements for the bicycles.

EXAMPLE

- The customer prefers to buy bicycles from a company in its own geographic area, because it wants to help the local economy and thinks that having a local manufacturer will minimize inventory and delivery problems.

  STRATEGY: Emphasize your long-term presence in the community and the accessibility of your manufacturing and storage facilities to the customer. Also, cite production contracts with other area customers and note that, for this contract, you'll use only subcontractors located close to the customer.

■ *Strategy Step #3: Conduct an objective self-assessment and make strategies based on your strengths and weaknesses.*

List the strengths and weaknesses in your ability to provide the bicycles. Then translate your strengths into (1) major selling points (themes) and (2) unique themes that differentiate you from your competition (known as discriminators). Finally, develop strategies that will shore up your weaknesses or at the least mitigate them.

EXAMPLE

- Your company has never missed a delivery deadline for its bicycles and has won quality awards for bicycle reliability and maintainability.

  STRATEGY: Translate these strengths to a major proposal theme—you provide proven on-time production and delivery of high quality bicycles.

- Your customer desires a personal and reassuring relationship with experienced contractor management; however, your long-time production supervisor will retire shortly after contract production is to begin.

  STRATEGY: Offset this weakness by committing the presence of the supervisor through the first six months of the contract to ensure a smooth implementation of the contract. In addition, show management continuity by highlighting the experience and qualifications of the assistant supervisor, who will replace the retiring supervisor.

Figure 1–4 shows how themes of your company and competition are defined. In your strategy planning, work hard to identify your discriminators, for they are themes that separate you from your competition.

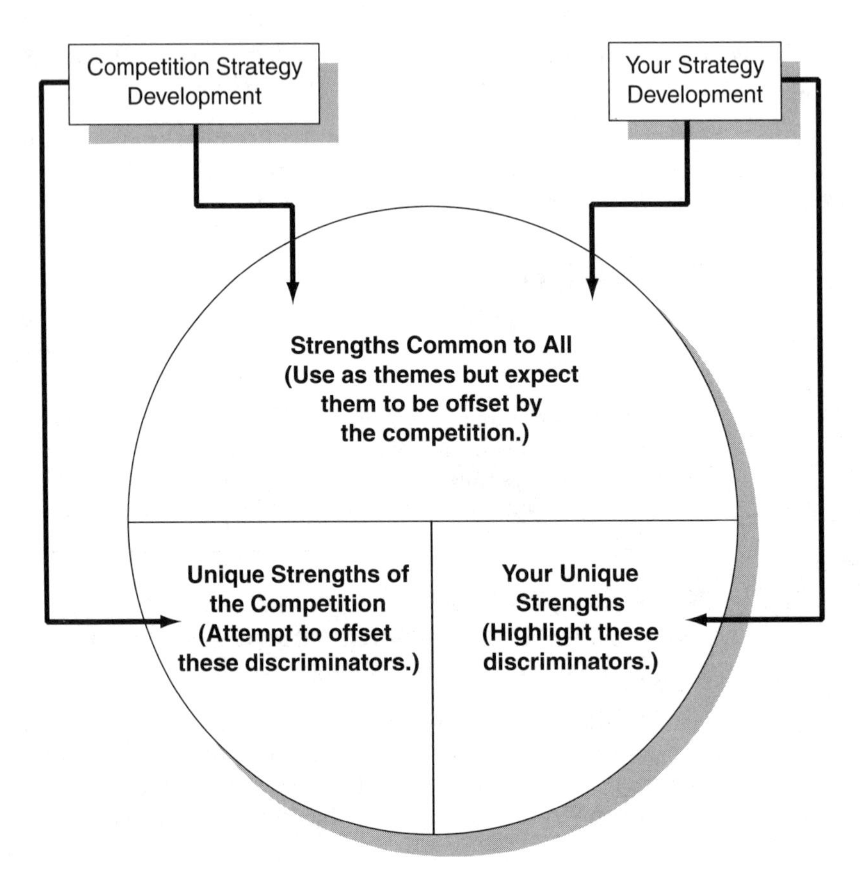

FIGURE 1–4
Themes for you and your competition

■ *Strategy Step #4: Conduct an objective competitor assessment and make strategies based on competitor strengths and weaknesses.*

> To *ensure* you win, your strategy must address your competition as much as it does the requirements, because the customer's source selection process ultimately *compares* the bidders and selects the best. So even if your proposal offers great benefits against each important requirement, if a competitor offers *better* benefits, you will have spent all that time and money in vain. The only way to ensure that you will win is to make sure . . . your most important benefits are better than the competition's.
>
> (Hansen, 1992, p. 146)

List the strengths and weaknesses of all your competitors in their ability to provide the bicycles. Then develop strategies to offset their strengths. If your competition has weaknesses, develop "ghosting" strategies to highlight those weaknesses by emphasizing your strengths without identifying your competitors by name.

EXAMPLES

- Your competitor's bicycle meets or exceeds the customer minimum standards, while your bicycle just meets the standards. However, based on your marketing intelligence, you believe that the competitor's model is 20 percent more expensive.

  STRATEGY: Emphasize that your bicycle economically complies with all customer requirements; it doesn't provide the customer with unwanted or unnecessary features.

- You also observe that your competitor has a unionized labor force that has gone on strike twice in the past 10 years.

  STRATEGY: Without mentioning your competitor by name, you "ghost" this weakness by emphasizing that your union-represented labor force has never conducted a strike against your company during its 20-year history, and that your management and the union maintain a strong supportive relationship.

Also determine if any of the competitors (including yourself) are providing bicycles to this customer. (A company currently under contract and bidding on the new contract is called an incumbent.) If there is an incumbent, it can indicate that the customer isn't pleased with the bicycles it has received from that contractor. If your competition is the incumbent, look closely at why the customer may want to rebid the contract and develop strategies to capitalize on the incumbent's problems. If you're the incumbent, take an objective look at your current performance and customer relationship, and develop strategies to overcome any real or perceived problems.

Being an incumbent can have its competitive advantages: (1) you already know the customer, (2) you should understand the technical and management demands of the job, and, (3) because you have contact with the customer, you might have the leverage to help the customer produce an RFP more in tune with your capabilities. However, being an incumbent can also lead you to be overconfident in your ability to win the new contract, resulting in a less-than-dedicated effort in producing the proposal. Plus, as noted previously, a new solicitation can mean the customer is displeased with your current performance.

■ *Strategy Step #5: Document your strategy meeting and develop a written strategy plan.*

Have one person, such as a proposal manager or project manager, run the strategy meeting. During the session, have this facilitator document the discussion to help the group brainstorm and organize its thoughts and to provide a record so that notes can be translated into a written strategy plan. Consider assigning an observer or a participant to also take notes and back up the facilitator.

Based on the strategy meeting, document the discussion and resulting strategies in a strategy plan book. Organize the book in sections to allow them to be cross-referenced to where they will affect the content of specific proposal sections. Figure 1–5 shows how the results of the strategy meeting can be translated into a strategy plan document that can be used by the proposal team.

As shown in Figure 1–5, the example strategy plan is written with five major sections. These sections include the following information:

- Section 1—your proposal themes based on your strengths, discriminators, and winning strategies
- Section 2—key customer concerns, issues, risks, and requirements, as well as your resulting strategies
- Section 3—your weaknesses and your resulting strategies
- Section 4—your concerns, issues, and risks, as well as your resulting strategies
- Section 5—competitor strengths and weaknesses and your resulting strategies

Number your strategies and themes based on this heading format.

EXAMPLE

STRATEGY: In Section 1, you list five proposal themes (key selling points) as Sections 1.1–1.5. The first two themes are the following:

- Theme 1.1—a management theme about the innovative way the management staff of your bicycle manufacturing department is organized
- Theme 1.2—a technical theme about the demonstrated quality and reliability of your bicycles

Strategy Step #6: Implement your strategy plan into your proposal through a requirements matrix.

Using the structure and format of the strategy plan headings, cross-reference strategies and themes to the proposal sections where they need to be addressed. This cross-referencing is made in the requirements matrix, which also indicates where RFP requirement will be addressed in the proposal. More information about the development and use of the requirements matrix is in Chapter 3.

Strategy Step #7: Update your strategy plan and distribute changes to the proposal team.

Update the strategy plan to reflect strategy or theme changes when new concerns, issues, strengths, requirements, or weaknesses are identified. If the strategy plan was developed before RFP release, conduct a follow-up strategy meeting and update the plan based on an analysis of the RFP after it arrives. Update and distribute changes in the strategy plan to all members of the proposal team and update

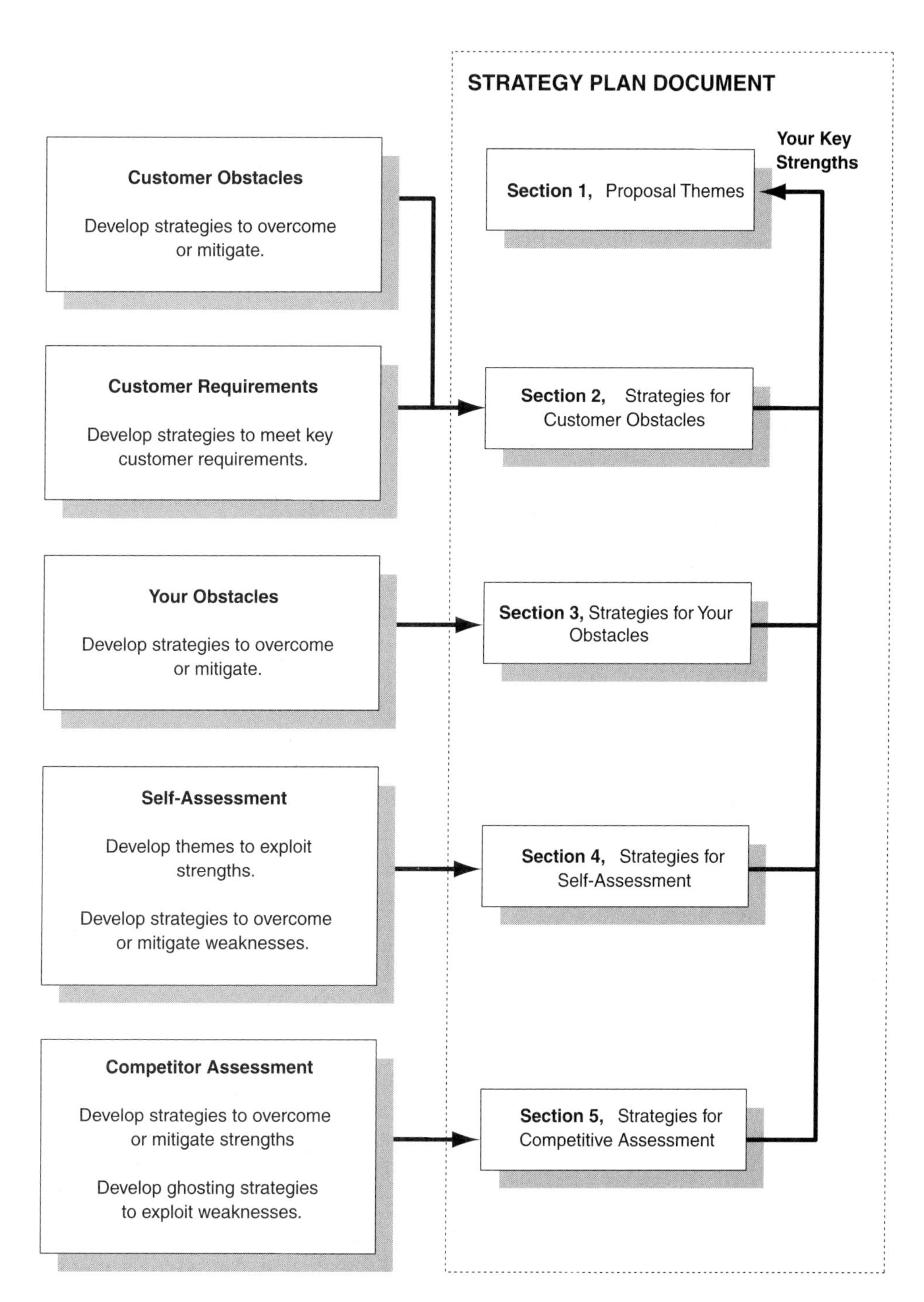

FIGURE 1–5
Strategy planning outputs for the strategy plan document

the requirements matrix if any cross-references to the strategy plan must be added or changed.

> Effective capture plans are based on accurate intelligence about the market, your customer, and your competitors. After making a pursuit decision, you will need to gather additional intelligence and continue to position your organization with the customer as you draft and update your capture plan. Gathering intelligence and positioning continue throughout the sales process.
>
> (Newman, 1994, p. 4-1)

SUMMARY

Proposal writing is only part of the larger goal of building trusting relationships with customers and managers. You can establish this trust through a series of repeated and helpful contacts. For external sales proposals, these contacts can include these steps:

Stage #1: Breaking In

1. Send a letter to a prospective customer.
2. Telephone the person to whom you send the letter.
3. Send a letter immediately after the phone call and before your meeting.

Stage #2: Getting Known

4. Meet with the customer.
5. Send a letter immediately after the meeting.
6. Influence the solicitation and qualify to receive it.
7. Write a dynamic, persuasive, and responsive proposal.

Stage #3: Clinching the Job

8. Provide required written and oral information after proposal submittal.
9. Request a debrief and write a follow-up letter after the proposal decision.
10. Negotiate by searching for shared interests.

Stage #4: Following Through

11. Submit timely status reports.
12. Send a letter of thanks with the invoice and follow with a "refresher" letter.

For in-house proposals, follow these steps:

1. Meet with your boss.
2. Write a persuasive proposal.
3. Make an oral presentation.

4. Complete a feasibility study.
5. Evaluate the completed project.

Justify the commitment of proposal resources by conducting a formal bid/no bid process before starting the proposal.

To develop winning strategies for your proposal, follow these steps:

1. Identify your key obstacles and make strategies to overcome them.
2. Identify key customer needs and make strategies to meet them.
3. Conduct an objective self-assessment and make strategies based on your strengths and weaknesses.
4. Conduct an objective competitor assessment and make strategies based on competitor strengths and weaknesses.
5. Document your strategy meeting and develop a written strategy plan.
6. Implement your strategy plan into your proposal through a requirements matrix.
7. Update your strategy plan and distribute changes to the proposal team.

EXERCISES

The following exercises can be the basis for a group discussion, an individual oral report, or an individual written report.

1. Sales proposal process: Real content.

Use nonproprietary company data from your own or others' work experience. Assume that (1) within the past year you decided to market a product or service to a customer with whom you've never done business, and (2) you were successful in getting work with this new customer. Now report on how you would have worked through the 12-step process for sales proposals outlined in this chapter. Be specific about your objectives and techniques at each step of the process.

2. Sales proposal process: Simulated content.

You are responsible for new-product marketing at a midsized firm in West Paris, Maine, that manufactures safety equipment. Your firm recently designed and began producing a new type of plastic safety goggles. A special coating, applied during the manufacturing process, radically minimizes the possibility of the goggles fogging up on the inside. (Assume that fogging is a common problem in safety goggles because of perspiration and because of many manufacturing conditions in which goggles are used.) Your firm wants you to market the new goggles to a large manufacturer of lawn mowers that has several plants throughout New England. Your contact is Ms. Sharon Shapiro, vice president of manufacturing at Lawn Helper Products of Hartford, Connecticut.

Using this fictitious case, and inventing additional details when necessary, report on how you would work through the 12-step sales proposal process outlined in this chapter. Be specific about your objectives and techniques at each step of the process.

3. In-house proposal process: Real content.

For the context of this exercise, use any organization to which you now belong—your employer, a civic group, or a college or university, for example. Select a nonproprietary idea that you could conceivably propose to the decision makers in this organization. Now report on how you would complete the five-step process for in-house proposals outlined in this chapter. Be specific about your objectives and techniques at each step of the process.

4. In-house proposal process: Simulated content.

You are an engineer at a small firm that designs and installs heating, ventilating, and air conditioning (HVAC) systems in new commercial buildings. You and 20 engineer colleagues write proposals for business, as well as project reports, to your customers after each job is completed. Lately, you seem to be spending about 75 percent of your time writing and polishing documents, taking too much time away from your technical work. Even then, you sometimes wonder about the quality of the final drafts. There's no in-house technical writer or editor, so it's up to the engineers and word-processing typists to catch errors.

You are convinced that hiring a technical writer/editor would save the firm considerable money. Engineers would continue to write drafts, but the writer/editor could help by (1) suggesting organizational improvements, (2) making prose style clearer and more appropriate for readers, and (3) working with the typists to eliminate mechanical and grammatical errors.

Your major obstacle is the strong opinion of the manager of engineering: you have heard him declare that all engineers should be expert writers and, therefore, responsible for their own reports and proposals. Given this obstacle, report on how you would complete this chapter's five-step process for in-house proposals. Be specific about your objectives and techniques at each step of the process.

5. Bid/no bid and strategy process: Simulated content.

Use nonproprietary company data based on your own or others' work experience. Your company has received an RFP solicitation to provide a product or service for a potential customer who has never done business with your company. The RFP was received unexpectedly by one of your sales representatives, whose only prior contact with the customer was a brief conversation with the customer's vice president at an industry trade show. You determine that your company can provide the requested product or service. Before committing the resources to submit the proposal, your manager asks you to determine if your company should bid this proposal. Identify the business, cost, schedule, technical, and management factors that could influence this bid/no bid decision. For any perceived weaknesses, concerns, or issues that you identify in this analysis, offer strategies that would resolve or mitigate them and support a bid decision.

Proposal Management

OBJECTIVES

- **Learn about the functions and organization approaches that support the proposal process**
- **Understand the importance of having a documented and repeatable proposal process**
- **Learn about sources for proposal training**
- **Learn to schedule and track the progress of the proposal process**
- **Learn the importance and details of 10 plans that help you manage the proposal process**
- **Learn the benefits of meetings and how to run meetings more effectively**
- **Perform exercises using chapter guidelines**

If your proposal management approach varies according to the people who run it, the quality of your proposals can vary greatly. To avoid this problem, use a proposal management structure with defined responsibilities and clear lines of authority, and ensure that proposal management follows a documented and repeatable proposal process.

In this chapter, we'll describe approaches to proposal management, including the features of proposal teams, and the tools and processes they should use for producing winning proposals. In today's business world, it's rare for a proposal to be the product of only one person. Therefore, our description of proposal management focuses on team proposals. With tailoring, however, team management procedures can be applied to solo proposal projects.

> Establishing an internal management plan for proposal development and production is key to long-term success. If you are going to stay in the business of preparing proposals, be serious about it. Take a long-term view and settle for nothing less than the best product you can produce.
>
> (Nocerino, 1993, p. xx)

Let's first identify the functions of a proposal team and how proposal teams can be organized. Then we'll look at proposal management tools and procedures.

PROPOSAL FUNCTIONS

The following are jobs that represent typical proposal development roles. The job titles can vary greatly. For a multivolume proposal supported by a large team with many writers, these jobs could be assigned among many individuals. For a small proposal and proposal team, one person might have several jobs.

- Proposal manager—as the proposal team leader, obtains all resources to produce the proposal, ensures the team meets the proposal preparation schedule, and leads the planning, development, and production of the proposal. However, these roles can be shared by two people: a project manager and proposal development manager. Such an arrangement allows a company to pair a proposal specialist with a technical/management leader who may have little or no proposal experience. In this relationship, the project manager can direct the technical, management, and price content of the proposal, and is a likely candidate to manage the proposed contract work if the proposal wins. The proposal development manager can support the project manager by (1) organizing and managing the proposal team, (2) procuring proposal materials, (3) planning and tracking the proposal preparation schedule, (4) developing the proposal outline and requirements matrix, and, (5) for smaller proposals, editing the proposal.
- Volume leader—leads the development of a proposal volume for a multivolume proposal, and reports to the proposal manager. The role is similar to that of the proposal manager, except that it focuses on a specific volume. Large proposals often have volumes for the technical approach, management approach, related experience (your current or recent contract work), and proposed price. The volume leaders, who work with their assigned volume authors to comply with the outline and requirements matrix, can be proposal specialists or be chosen because of their technical or management skills for the volume subject. They may also edit their respective volume.

- Proposal coordinator—helps the proposal manager control the proposal development process. This job can include almost any proposal development task. For example, the coordinator may (1) update and distribute the proposal preparation schedule, (2) copy and distribute RFPs and other documents, (3) manage text and graphic files produced by the team, (4) prepare and copy proposal drafts for review, and (5) obtain and distribute proposal office supplies.
- Production leader—leads the production (publication) process, including graphic and word processing support, text and art integration in the proposal layout, and the printing of the proposal document. These tasks also may be supported by the proposal coordinator. The production job can be performed by a company organization that not only supports proposals but also serves publication and media needs for product manuals, technical reports, and marketing presentations and brochures.
- Proposal editor—edits proposal drafts. As previously noted, this role might be assigned to the proposal manager or volume leader. However, these people don't always have the time or skill to be a good proposal editor. Unfortunately, whether it's the function of a dedicated editor or the collateral job for others, editing often receives little priority or time in the proposal development schedule. Adequate resources and time must be committed for this crucial part of the writing process. Chapter 10 describes editing techniques.
- Proposal writers and illustrators—work for the proposal manager or volume leaders to create text and graphics. The writers can be full-time proposal specialists or individuals "borrowed" from functional departments in the company. With today's software, many proposal writers produce their own text and graphics. However, it's common for larger proposals to have a graphics group produce the graphics, based on information from the proposal writers. In other arrangements, an illustrator can be assigned to help authors conceptualize and draft their own graphics.
- Quote and pricing staffs—develop the price of the proposed product or service for the cost portion of the proposal. The roles of people and organizations involved in this pricing effort depend on what is being offered and the company's policies and procedures. Pricing can be a simple process if the customer is buying off-the-shelf products with a catalog price. However, the more a product or service is tailored for a customer, the more complex you can expect the quoting to be. (Quoting is estimating that forms the basis of the proposed price.) For example, the price of developing a new product might have to reflect labor hours, material, and travel costs for engineering, testing, manufacturing, and project management activities. In addition, you would expect to have overhead and a profit margin added to these costs. Although the focus of this book is on the written proposal document, not on pricing the product or service, it's important to recognize quote and pricing staffs as performing key proposal functions.

PROPOSAL ORGANIZATIONS

> Any proposal manager (or outside consultant, for that matter) who claims to know how to run a serious proposal effort without chaos should be quietly and briskly dismissed from reality. Chaos is the very nature of the proposal beast, and the manager who accepts this fact is ready to face another one: While the chaos of proposal work cannot be eliminated, it must be controlled. Otherwise, it will eat the manager, the proposal team, and the proposal itself alive.
>
> (Pugh, 1993, p. 82)

There are many ways to organize the preceding roles in your proposal group. The structure and composition of the proposal organization depend on the following factors:

- The size, organization, and staff skills of your company
- Your budget for bid and proposal (B&P) funds, which pay the material and labor costs for developing proposals
- The complexity and turnaround time of your proposals and their page count
- Your company's business volume and the number of proposals you produce
- Teaming arrangements you have with other companies, subcontractors, and vendors to produce proposals
- Technology used to perform communication, publishing, writing, illustration, information storage and access, proposal management, and pricing

Figure 2–1 shows the possible organizational structure of small and large proposal teams, and includes the functional positions described in the previous section.

Control of Proposal Development Costs

The peaks and valleys so common in the proposal workload of a company make it difficult to predict the optimum size of a proposal staff. With staffing at a set level, there can be so much work that just delivering the proposal on time—regardless of its quality—is an achievement. At other times, the lack of proposal work can cause slack periods, leading to questions about whether the proposal staff is too large or, indeed, whether a separate proposal staff is even justified. The labor involved in proposal work is probably the biggest part of a company's B&P costs. To control these costs, you must make your proposal labor as productive as possible.

Use judgment in how big the proposal staff is and when it works. Although early proposal planning and preparation can pay dividends, don't commit too many people too early in the process, especially if you're waiting for the release of an RFP. If you have a limited B&P budget, and the RFP release date slips, a large planning proposal staff can quickly exhaust much of the budget before the RFP is received. The solution can be a small core group of key personnel assigned before the RFP is released. For example, the core group—which might include the pro-

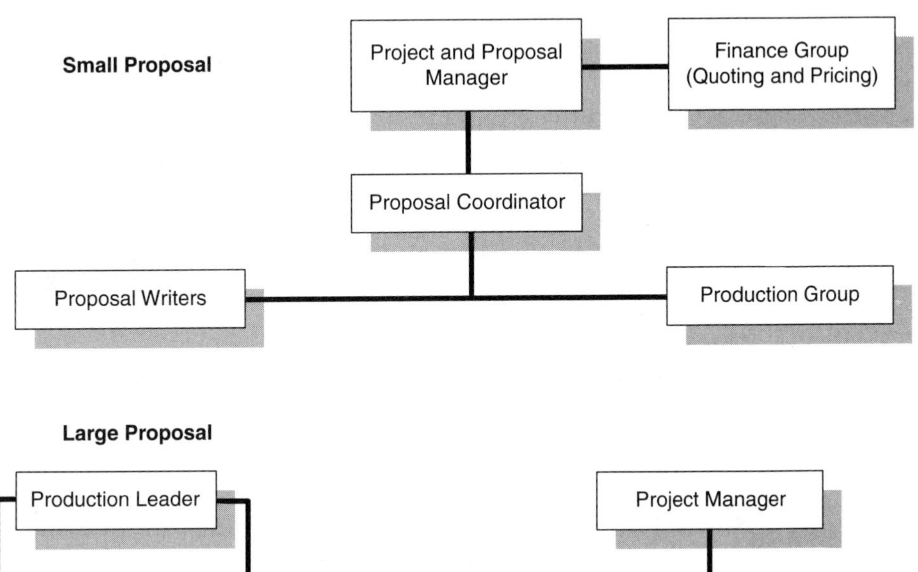

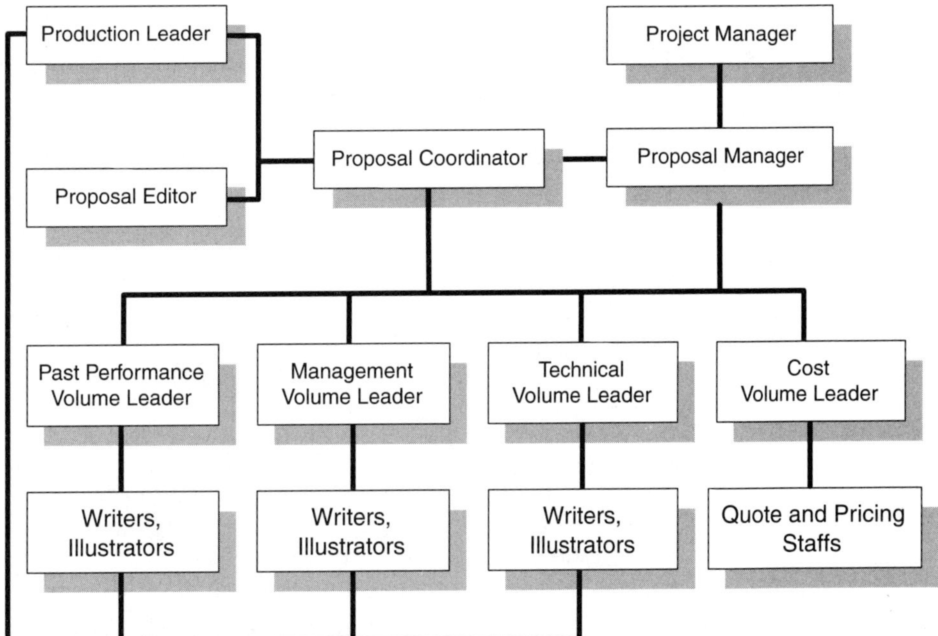

FIGURE 2–1
Possible approaches for organizing proposal teams

posal manager, project manager, and a technical lead—could be assigned only long enough to analyze and comment on the draft RFP, conduct a strategy meeting, and develop technical, management, and cost baselines before RFP release. When the RFP is received, the core group then could be reassigned to the proposal and serve as the nucleus of the full team.

Managing your labor resources isn't just a pre-RFP concern. Always assign personnel to a proposal only when they're needed. This practice can save you labor costs and reduce the risk of having people in the proposal team area who are idle and become distractions to those who are working. If you do a good job of staff and proposal management, the size of your team should decrease as the proposal matures, because, as it develops, the proposal should require less labor to be ready for delivery. For example, you should be able to trim the size of the proposal staff after the final draft is complete and then turned in for final editing and production. Unfortunately, a proposal team can find itself so far behind in terms of schedule and proposal quality that the staff size must increase late in the proposal process to overcome proposal deficiencies.

To also help reduce proposal labor costs, consider the following suggestions:

- Provide training to your staff so it can perform more efficiently.
- Establish standard procedures to avoid developing new procedures for each proposal.
- Plan before the writing begins to avoid rewrites (and restarts) caused by a team that doesn't know what it's supposed to write.
- Use the most cost-effective mix of full-time and contract/consultant employees.
- Implement technology that helps your staff to more quickly and effectively do its job.

Organization and Staffing Approaches

The following are possible approaches for organizing and staffing proposal teams. If a good proposal process is followed, each approach can be effective. Table 2–1 lists the major advantages and disadvantages of each approach, including labor cost considerations.

- Dedicated—This organization is committed solely to proposal development with a group of full-time proposal specialists assigned to perform all or some of the preceding proposal functions. Its array of services can vary from a full-service approach in which all resources are provided by the group to a more specialized service, such as proposal management, which directs writer, graphics, and production personnel from other company groups to produce the proposal.
- Collateral—Instead of a dedicated proposal organization, proposal development is performed as a collateral job by a functional group, such as a marketing, sales, or publications department. Like a dedicated proposal organization, a collateral organization can produce the proposal by itself or direct a proposal team consisting of people from different functional groups in the company.
- Project—The proposal team is staffed by the people who are responsible for delivering the product or service offered in the proposal. For example, the expected project manager of a construction project might form and lead the proposal team consisting of architect, engineering, construction, manufacturing, and environmental personnel who will design and perform the construction.
- Outsource—With little or no internal proposal resources, a company hires consultants or contractors to produce the proposal with company support.

TABLE 2–1: Advantages and Disadvantages of Different Proposal Organization Approaches

Type	Advantage	Disadvantage
Dedicated	• The group supplies a cadre of skilled proposal professionals to provide consistent and quality proposal development and management. • Group resources can be shared or distributed among several ongoing proposal projects.	• It can require a significant financial commitment with nondirect funding. • It can be difficult to handle high levels of proposal activity with limited resources or to justify the expense of maintaining these resources during low levels of activity. • Proposal staff understands proposal techniques but may have weak technical knowledge of the proposed product or service.
Collateral	• The proposal team is formed only when needed, avoiding the cost of maintaining a dedicated proposal group. • Frequent involvement in the proposal process by the same organizations can lead to strong and experienced proposal management, even if it's a part-time responsibility.	• Proposal staff has to balance its commitment between proposal and regular job responsibilities. • Proposal staff may have little proposal experience and proposal training. • People can be assigned to the proposal staff not because of their skill but because they are available, regardless of their skill or personal commitment.
Project	• Knowledgeable and motivated people are brought to the proposal staff. They should understand the proposed product or service and be inspired to succeed, because, if the proposal wins, the staff gains future work. • The proposal team is formed only when needed, avoiding the cost of maintaining a dedicated proposal group.	• Staff may be technically proficient in the product or service but have limited proposal skills, with little proposal experience and proposal training.
Outsource	• External sources provide skilled proposal professionals who can bring new and objective ideas for proposal processes and content. • External sources can supplement internal proposal capability and resources when the support is needed the most. • In the long term, it can be less expensive to rely on external experts than it is to sustain internal proposal resources.	• Consultants and contractors require time to learn about the procedures and personal dynamics of the company and its products and services, as well as to establish rapport and trust with company employees. • Outsource hiring can be very expensive in the short term. • It can be difficult to screen and assess the quality of consultants and contractors before hiring them. • Consultants and contractors may not be available when you need them because of their other commitments.
Hybrid	• The integrated approach can build an "all-star" proposal team with the best collection of internal and external proposal resources.	• The many sources of proposal staff resources can lead to great variations in processes, techniques, staff ability, and proposal quality, making it harder to integrate and manage the proposal team. • The approach can be confronted with all the problems associated with the other types of proposal organizations.

Although retaining responsibility for proposal content, the company hires them to collect the required information and then to write and prepare the proposal. In less inclusive arrangements, the company can outsource specific functions to supplement its own proposal resources or to provide services that it can't provide for itself.

- Hybrid—Like the answer to a multiple-choice question—"all of the above"—this approach is a mix of the preceding approaches. It can result in many different types of proposal teams. For example, the company might form a proposal team with proposal specialist, marketing, engineering, and selected contract/consultant resources, led by a project manager, proposal specialist manager, or a proposal consultant.

PROCESS DEFINITION, TRAINING, AND SCHEDULING

> There are as many ways to write proposals as there are companies, but many of these ways are crude, ineffective, expensive, and bound to repeat many previous mistakes. . . . We cannot stress enough that organized proposals are probably twice as likely to succeed as those "fly by the seat of the pants" exercises. . . .
>
> (Whalen, 1996, p. 1-3)

All proposal efforts should include (1) a documented process, (2) the training to teach the process, and (3) a schedule to implement the process.

The Documented Process

Define your proposal process in writing. It can be a summary process, providing only basic guidelines to be tailored for each proposal, or a very detailed checklist used for every proposal. A documented and enforced process helps avoid "reinventing the wheel" each time a proposal is tackled. It can also set standards in your process, promoting at least a minimum—and repeatable—level of proposal quality regardless of the experience of the assigned staff. As a process, it can be improved based on lessons learned in actual proposal work.

Ensure that the process guidelines are available to the proposal team. You can distribute them in a procedures and information book to the team at the proposal kickoff meeting. Often called a proposal directive, the book can contain many details, including the following:

- Proposal preparation schedule
- Detailed instructions and responsibilities for each scheduled proposal task
- Planned proposal strategies and themes
- Summary of the baseline approach for providing the proposed product or service

- Writing and graphic development tips
- Style and format of the proposal content
- Summary outline and requirements matrix
- Organization and roles of the proposal staff
- On-line and off-line work flow and communication procedures

The book should be updated as required by proposal management during the proposal project.

Also consider developing a "how-to" proposal management guidebook to help teams without experienced proposal management. This guide can present the most pertinent instructions from the proposal directive in an abbreviated checklist and reference format. For example, it could be especially valuable for the development of small engineering research and development proposals by engineers who are assigned to lead their first proposal effort.

Training

For a company to implement the proposal process, its proposal staff must receive proper training. It can be taught by company proposal specialists, external proposal training firms and consultants, and college/university degree and continuing education programs. Training can be provided when participants are between or on proposal assignments.

Company instructors and "students" may be more readily available if the training is given between proposal assignments. This training can be short or long formal instruction. For example, a series of company-provided training classes packaged in a comprehensive proposal curriculum could be offered at various times. In a longer, yet accelerated, company training course, employees could participate in a team workshop in which they plan and produce a proposal based on a fictitious RFP or a real RFP that was once responded to by the company. Company proposal specialists could serve as proposal manager instructors for the team, critique the proposal, and provide feedback to the team about its performance. A company can also have employees formally trained by outside training services. This instruction could be held at a facility provided by the company or training service.

However, if the trained skills are not used "under fire" until long after the training is given, the lessons may be forgotten when they're needed the most. To avoid this problem, the company can use what is known as "just-in-time" training, which is held for employees working on a proposal project just before the applicable skills are needed. For example, the proposal team might attend a half-hour class about storyboard development just before beginning that task. After finishing the storyboards, the team might then receive a short primer on persuasive writing before starting the first proposal draft. Although this training could be provided by the company or outside training service, it might be best to hold such short-term training at the company's facility. While working on a proposal, a team can also use the proposal directive as a proposal training reference.

Scheduling

Even with a defined process, writing a proposal without a schedule is like driving to a new destination without a road map. If you don't use a schedule, it's difficult to judge where you're going and what progress you're making to get there. The following are tips for making a schedule to prepare a proposal.

■ *Schedule Tip #1: Wisely use the time allowed for proposal preparation.*

The amount of time it takes to complete a proposal depends on its delivery deadline, the resources you can commit to the proposal, and how well you manage those resources.

Look for ways to start your proposal even before the RFP is released. While awaiting the RFP, schedule and perform tasks that you know must be met, even if you don't know what the RFP will require. If you do have a good idea what will be in the RFP because you have a draft RFP or have made frequent pre-RFP contact with the customer, you might plan a schedule to write at least a first proposal draft based on this advance information.

Be aware of tasks that require a long lead time before they can be finished. For example, if you need pricing information from subcontractors and vendors, begin this task early by meeting with them in a quote coordination meeting and providing them with a preliminary statement of work for their involvement in the proposed contract. By identifying key prerequisites, you can focus early on these tasks to ensure they are completed on time.

Submitting your proposal much earlier than required might seem like a good idea. However, customers don't normally give bonus points for early submittal, and the unused proposal time could limit your proposal's effectiveness—especially if your competition uses all of its allowable time well. Submitting your proposal early or even printing the final draft early and holding it for delivery just before the deadline can also lead to problems. If the customer makes last-minute changes to the RFP, it could require proposal content changes and force a new round of proposal drafts, reviews, and printing.

Use your time wisely at the end of the proposal schedule. Last-minute editing and re-writes to make the "perfect" proposal have diminishing returns and can delay proposal delivery. Stay focused on meeting your schedule, know when to stop working, and admit to yourself that, although a perfect proposal may be a worthy goal, it's doubtful that one has ever been written.

■ *Schedule Tip #2: Use task- and responsibility-based schedules driven by your proposal management plans.*

Schedule tasks drawn from the 10 proposal management plans described later in this chapter. Clearly define the tasks and identify the people responsible for each scheduled milestone to complete these tasks.

■ *Schedule Tip #3: Schedule a seven-day workweek only when it's necessary.*

In a perfect world, all proposal team members would dedicate the time needed to meet their assignments without a mandated presence at their proposal desks. However, reality dictates the use of a schedule for discipline. Set a daily and weekly schedule that, while it meets the demands of the proposal, still allows ample time off for the team. Although a seven-day workweek may seem necessary at first, the intensity of this schedule can lead to team burnout and reduced flexibility in meeting unexpected problems. If you fall behind schedule with a seven-day workweek, you might work longer hours or add more people, but you can't extend the workweek to eight days. A seven-day schedule can actually result in less productivity, because it may encourage the team to take the attitude of "Why should I do this now when I have the weekend to do it?" Don't have unnecessary people assigned to the proposal team or in the proposal work area. As previously noted, having only the required people assigned to the proposal can save money and reduce personal distractions.

■ *Schedule Tip #4: Avoid short-changing scheduled tasks performed early and late in the proposal process.*

Schedule enough time to plan and outline the proposal early in the process and to edit, lay out, and print the proposal at the end. Expect these tasks to require more time than you think they should. Fight the temptation to delay the completion of unpleasant or difficult tasks. If you defer too many tasks, the later schedule may become so full, you may never finish them. Set a firm, nonnegotiable deadline for the completion of all final drafts. This approach can help you avoid the common problem of writing, review, and revision tasks encroaching into time for proposal production.

■ *Schedule Tip #5: Be sensitive to what day of the week you schedule a milestone.*

Not all days are created equal. When you make your proposal preparation schedule, think about how the day of the week will affect the completion of the milestone. For example, it might be best to schedule a draft input for Friday, because it gives authors an incentive to complete it before the weekend. In another example, your first inclination might be to schedule a review and its debrief for Friday and then have the proposal team take off on the weekend and return on Monday to begin the incorporation of review comments. Instead, it might be more productive to schedule that review earlier in the workweek to allow a quick and focused response to these comments while the review results are still fresh in the minds of the proposal team. With the review in the middle of the week, it might be easier to get reviewers—especially those from out of town—to come in the day after the review to help the proposal team.

■ *Schedule Tip #6: Look for opportunities to divide major task milestones into staggered, smaller tasks.*

Avoid taking an all-or-nothing approach to completing milestones. For example, instead of scheduling the final draft input of all proposal sections on the same day, you might avoid production bottlenecks by staggering the input of final section drafts. This staggered schedule could allow a shorthanded or busy production group to work only the inputs it can handle and to give some sections more development time before they enter final production and edit. You might also stagger the reviews of different volumes or major sections on different days, distributing the workload of the review team over several days, accommodating the availability of a needed reviewer, or giving the proposal team extra time to complete a troublesome volume before it's reviewed.

■ *Schedule Tip #7: Monitor schedule progress, maintaining a balance of holding firm to a schedule and changing it when needed.*

Regularly review the progress of the proposal preparation schedule, identifying and resolving problems before they become serious. Stress compliance with the milestone dates, but make sure the milestones are based on a realistic schedule. Have enough flexibility to change the schedule if required. Distribute updated copies of the schedule to the team, indicating milestone status and any revisions that have occurred. Provide the schedule and updates to all proposal participants, including supporting groups, such as internal marketing, finance, business development, sales, and production groups, and external teaming companies, subcontractors, and vendors. A good schedule reflects that proposal management must rank the priority of its tasks. If the proposal can be competitive without completing a task, maybe that task shouldn't be scheduled.

THE PROPOSAL MANAGEMENT PROCESS

> . . . managing a proposal is one of the toughest, most challenging jobs you will ever encounter. It involves dealing with a wide variety of people with conflicting interests and organizing and directing the efforts of an *ad hoc* collection of people with diverse talents and expertise into a cohesive, motivated *team* under the most adverse circumstances in an intensive activity and under great pressure to achieve a goal against which the odds of success are not favorable. Can you think of a more daunting endeavor anywhere else in the world?
>
> (Helgeson, 1994, p. 206)

Use the following 10 steps to manage your proposal process. These steps represent plans that should be reflected in your documented proposal process, training programs, and proposal schedules. Table 2–2 summarizes the tasks for each plan.

TABLE 2–2: Proposal Management Plans

Plan	Typical Proposal Tasks
Capture plan	Performing customer analysis, strategy analysis, and lessons-learned analysis
Customer interface plan	Contacting the customer before RFP receipt, during proposal development, and after proposal delivery
Facility and material plan	Obtaining facility space and furniture, office equipment, general office supplies, and reference documents
Staff plan	Organizing, staffing, and managing the proposal team
Communication plan	Controlling team communication, including the use of communication tools, schedules, meetings, and reports
Document process plan	Processing and controlling the configuration of storyboards, mock-ups, text, and graphics
Quality assurance plan	Reviewing and editing proposal drafts
Production plan	Preparing support material, setting guidelines for formatting and style, integrating text and graphics, and printing the proposal
Security plan	Safeguarding and controlling proposal information, including compliance with communication, material handling, and personnel security procedures
Assembly and delivery plan	Assembling, book and page checking, packaging, and delivering the proposal

Proposal Management Step #1: Develop and implement a capture plan.

The capture plan consists of (1) a customer analysis, (2) a strategy analysis, and (3) a lessons-learned analysis.

To respond to an RFP, analyze it carefully to determine customer requirements for proposal content, format, and organization. As will be described in Chapter 3, use the RFP to produce a summary outline and requirements matrix to ensure responsiveness to all customer requirements. The requirements matrix cross-references the proposal outline sections to the applicable RFP requirements they must address.

As fully described in Chapter 1, conduct an analysis to identify strategies that will boost the competitiveness of your proposal. Document your strategies and themes in a strategy plan for use by the proposal team. If the RFP is available during this strategy analysis, use it to help evaluate the customer needs and the ability of you and your competition to meet those needs.

As part of the strategy planning, draw on lessons learned in other proposal activity—whether it was successful or not. Avoid problems encountered in previous proposals and use the approaches that worked.

Proposal Management Step #2: Develop and implement a customer interface plan.

Plan how you will communicate with the customer before receiving the RFP, while developing the proposal, and after submitting the proposal. There are many opportunities to exchange information with the customer, including the following:

- Pre-RFP customer contact or briefings you initiate to learn about or influence the RFP requirements
- Pre- or post-RFP release briefings or meetings the customer initiates to gather industry comments about the requirements for the desired product or service, or about the content, format, and organization of the proposal
- Written responses you make to customer questions about your submitted proposal
- Physical demonstration of your product or service or customer visits to your facility as part of the proposal evaluation
- Oral presentations you provide to the customer about your proposed product or service after the written proposal is submitted
- Customer-initiated oral interviews and negotiations, or requests for your BAFO after the written proposal is submitted
- Customer debriefings after contract award—whether you win or lose—that can provide a valuable critique of your proposal

Whatever the customer contact, develop a plan about who will be responsible, and what, where, how, and when tasks will be done. It can be very easy to neglect customer contact that doesn't appear directly related to the production of a proposal document. If you don't communicate well with the customer during all contact opportunities, even the best written proposal can be a loser.

Proposal Management Step #3: Develop and implement a facility and material plan.

Obtain all the resources you'll need for the size of your proposal team and the scheduled duration of the proposal process, including the following:

- Facility—A dedicated proposal work area, always available for proposals or assigned only when needed, can boost team coordination, communication, and concentration. However, if one dedicated area isn't available, you'll need to have a plan to locate your proposal staff for team coordination and communication. Your facility choice can be affected by many factors, including the need for individual offices and desks for the proposal staff, telecommunication and data-processing lines, conference rooms, kitchen and restroom areas, and wall space to display drafts of the proposal. The sensitivity of the proposal material can also determine if a separate area must be assigned to the proposal team. Proposal Management Step #9 describes the security issue in more detail.

- Equipment—You may need computer hardware and software, printers, copiers, scanners, telephones, fax machines, audiovisual equipment for meetings, and creature comforts, such as a refrigerator or coffee pot. (Some may argue that the coffee pot is more than a creature comfort; it provides the fuel—along with a local pizza delivery restaurant—for driving the proposal team.) Your network approach for linking computer hardware and software can affect your facility choice and other hardware and software items you obtain. For example, your computer network, linking team members with each other and resource documentation, might be through a local area network or an intranet web site, each requiring hardware and software.
- General office supplies—Obtain an adequate supply of supplies, such as pens, pencils, erasers, felt tip markers, copy machine toner, paper (printer, copier, and fax), diskettes (and holders), folders, binders, and envelopes. Monitor and maintain your supply inventory and have these materials available with some control, perhaps with a supply cabinet accessible in the proposal area.
- Reference documentation—These materials can include copies of the RFP (including draft RFPs and amendments to the formal RFP), the strategy plan, proposal directive (including a proposal style guide), related proposals from the past, technical and design specifications, test and evaluation reports, standard management and technical plans, company management directives and procedures, and company press releases and news publications. Proposal teams can also have access to useful reference information from commercial and government Internet web sites and their own company-maintained databases and intranet web sites. Printed reference information can also be obtained from journals, newspapers, magazines, and a variety of U.S. government publications.

Proposal Management Step #4: Develop and implement a staff plan.

> Proposal work has an occupational hazard that does not exist in many other parts of industry today, and that is known as "burnout." Because good and willing proposal people are few and difficult to find, there is a tendency to assign too many projects to them at the same time. Working under the intense time and mental pressure for days, nights, and weekends can be taken by most people on an occasional basis, but when one has it for continuous fare, there comes a time when he or she just quits, either voluntarily or involuntarily, representing a dead loss to the organization.
>
> (Hill, 1993, p. 123)

Regardless of your proposal staff source—internal or external—organize the team with clear lines of responsibility and authority. It's crucial to permit the proposal manager and the assisting managers to lead and make firm decisions that will be implemented by the proposal team. People on a proposal team are often on loan from organizations in the company or from other companies involved as subcontractors, vendors, or joint venture members. These members may be more committed to their functional manager or company than to the proposal manager. To avoid this problem, proposal managers should be given the responsibility, authority, and resources to lead the proposal team and also receive strong support from organizations that loan people to the team.

As soon as possible, identify the proposal staff you will need from internal company resources and external contractors and consultants. (External contractors include personnel from subcontractor and vendor companies that will support the proposed product or service and contribute to the proposal.) This step helps you to obtain early commitment of full- or part-time staff; to begin early planning and preparation of the proposal; and to provide timely training for those who need it. This early start also allows time to find replacements for first-choice staff who are unavailable for team assignment or to arrange to have staff members—with otherwise conflicting work—assigned to the proposal as their top job priority. Early staff planning can also ensure the availability of contractors and consultants who must balance work commitments among various clients.

Once assigned, proposal staff members, including proposal management, must be held accountable for the quality of their work. If they don't perform as they should, the proposal manager must have the authority to replace weak performers if attempts to resolve the problem have failed. Once the decision is made to replace a proposal staff member, quickly make the switch, so that the replacement can become integrated into the proposal team.

Proposal Management Step #5: Develop and implement a communication plan.

Establish communication procedures for the team. Be responsive to the special needs of a proposal team that is dispersed in different locations. Develop procedures for using communication tools such as e-mail, Internet and intranet services, telecommunications, videoconferencing, voice mail, and fax machines. Implement procedures to keep the team aware of proposal information and progress with the following:

- Schedules—As previously described, regularly distribute an updated schedule of all proposal activity and include in the schedule who is responsible for the activity.
- Meetings—Schedule and conduct kickoff, status, and planning meetings. The kickoff meeting is held to mobilize and organize the proposal team and to inform it about the proposal management plans. Status and planning meetings are held for management of the entire proposal team or for volume teams. More details about holding effective meetings are presented later in this chapter.
- Status reports—Identify the frequency, content, and distribution of status reports to monitor the progress and cost of preparing the proposal. These reports help you to identify and resolve problems early and to operate within the B&P budget for your proposal project.

Proposal Management Step #6: Develop and implement a document process plan.

Set guidelines for controlling the flow and configuration of storyboards, mock-ups, text, and graphics. Flow procedures are used to control how the proposal product is processed through its various drafts. Configuration control ensures that the most current version of the proposal is safely stored and available to the team. The plan must include the following issues:

- Procedures for preparing, reviewing, and printing of proposal documents
- Management of computer files (soft) and printed copies (hard) of proposal documents, including the naming, storage, and backing up of soft files
- Distribution, use, and processing of hard and soft copy worksheets for making storyboards, mock-ups, and graphics, including tables and schedules
- On-line or printed copy access to the latest versions of the proposals, including the use of read-only computer files and wall-mounted hard copies of proposal drafts

Proposal Management Step #7: Develop and implement a quality assurance plan.

To produce a high quality proposal, it's important to dedicate enough time for thorough reviewing and editing. Identify the schedule, purpose, and needed resources for these functions. Because the review can be so crucial for a winning proposal, use this activity as one of the pivotal milestones in your schedule. Chapter 9 provides more details about reviews, while Chapter 10 describes editing.

Proposal Management Step #8: Develop and implement a production plan.

This plan addresses a broad range of proposal content, appearance, printing, and publication functions. They include the following:

- Support material—This material supports the main proposal sections. It includes the production of such items as the table of contents, foreword, title page, figure and acronym list, glossary, compliance matrix, cover letter, letter of transmittal, and section tab dividers.
- Format—Format involves the mechanical structure and appearance of the proposal. It includes the standards and procedures for proposal layout such as the spacing between text lines, column justification (block or ragged), number of text columns, size of the page margins, location of page headers and footers, and the integration of text and graphics on the page.
- Style—Style is the standard for the content and appearance of text and graphics. It includes the font style and type size of text, tables, captions, and graphic call-outs; the numbering style for text headings and graphic captions; and the capitalization, spelling, and content styles for captions, text headings, acronyms,

abbreviations, initialisms, titles, and special nomenclature. Careful editing should verify that style guidelines are being followed.

- Printing—Printing decisions include such issues as the number of proposal copies; the printing schedule, method, and source; the binding method; the type and color of paper; and any special provisions for the printing of foldouts, inserts, graphics, covers, and tabs. When feasible, do your proposal printing as early as possible. For example, the covers and tabs—if you're sure they won't change later—can be printed well before the main proposal. Proposal material that doesn't change from proposal to proposal can be printed before the proposal begins. For example, product pictures and specification lists that won't soon be revised, can be printed in quantity and stored for use by many proposal teams. Instead of printing copies of the finished proposal at one time, consider the incremental printing of proposal sections as they are completed. Avoid the problems with last-minute mechanical failures of equipment by always having a backup source for all printing needs.

Proposal Management Step #9: Develop and implement a security plan.

A security plan helps you comply with company and customer security regulations, including the care of your company's proprietary (competition-sensitive) data and U.S. government classified information. Security concerns can affect your plans to use facility and communication resources, to safeguard and dispose of proposal materials, and to store and access references. The plan can address such security issues as the following:

- Codes and passwords for access to the proposal areas, web sites, and server networks, as well as antivirus protection for software programs
- Personnel badges and parking passes to enter a company's facility
- Government-granted security clearances for those working on classified proposal projects
- Training to properly handle proprietary and classified materials
- Company physical and security force resources to permit the processing and storage of classified material
- Containers in the proposal area to dispose of proposal proprietary and classified material for controlled destruction
- Delivery methods and safeguards for classified proposals sent to U.S. government customers
- Customer security requirements that directly affect proposal content, such as proprietary data statements for noting competition-sensitive information and annotations for showing the classification level of the proposal information

Proprietary data statements, which describe the limits for customer use of the company's proposal information, merit more description. A common practice is to state the limitation on the title page and then, as a footer on each applicable page, to note that use of the proposal is restricted by the statement on the title page. The RFP may dictate the proprietary data statement to be used in the proposal. If it doesn't, use your own statement—cleared through your contracts and legal organization—that states the proprietary status of the proposal information and who can use it. (By the way, just because something is in your proposal doesn't make it proprietary data.)

Proposal Management Step #10: Develop and implement an assembly and delivery plan.

After printing the proposal document, assign people to assemble the proposal, including all pages, tabs, and binder covers (front, spine, and back, as required). For every copy that's going to the customer, do a thorough book and page check, verifying that all pages are inserted in order and printed with proper alignment and clarity. If you're using ring binders for holding the proposal, verify that the rings close properly to ensure that the proposal remains secured. When approaching your delivery deadline, concentrate on printing, assembling, and page checking only the proposal copies you need for the customer. Extra copies for internal use can be printed and assembled after the proposal is delivered.

Although the book assembly and page check session isn't the place for proposal editing and proofing, if major mistakes or typos are found, have a plan to make last-minute corrections. For example, you can have a disk copy of the proposal available with a computer and printer for correcting the mistake and then printing another copy of the affected proposal page.

Complying with any customer requirements, pack the proposal with the required packing material and containers. For example, you may need to submit a certain portion of the proposal in a separate box, have a specific identification label on the outside of the box, or address the delivery to a specific person. During shipment and delivery, good packaging can help you avoid damage to the binders and proposal pages or the slipping or tearing of pages from the binding.

Delivering your proposal one minute late can eliminate you from the competition. To avoid what can be a very expensive and embarrassing mistake, make primary and backup delivery plans to ensure you meet the delivery deadline. For example, even if your choice is overnight delivery by an always dependable shipper, you might have an alternate plan to hand carry the proposal by a staff member who is standing by with airline and rental car reservations if the proposal isn't delivered by a certain time. If you're shipping or mailing directly to the customer, consider sending another set of proposal copies to a trusted contact near the customer, who could hand deliver the proposal if it doesn't arrive by a certain time. Contact the customer to verify receipt of the proposal. If there appears to be a glitch in the primary delivery method, go to the backup plan. Of course, allow yourself enough time to implement the backup plan if it's needed.

If you travel by commercial air to hand deliver the proposal, avoid checking in the only copies of the proposal as check-in luggage or packages. If possible during the trip, hand carry all required copies of the proposal to ensure that you and your proposal go to the same place at the same time. For especially critical and large proposals, you might buy a separate airline seat to allow more cabin room for your valuable cargo. At a minimum, have at least one complete copy of the proposal in your possession at all times. This backup step allows you to deliver one copy to the customer even if other proposal copies in check-in luggage or packages are misplaced by the airline.

EFFECTIVE MEETINGS

> An enormous amount of time is wasted in aimless, rambling meetings replete with the musings and irrelevant chatter of unfocused dilettantes intent on wasting everyone's time while they run their mouths. I would wager that the manhours wasted every single day in the conference rooms of America would be equivalent to the number of manhours required to build the Taj Mahal.
>
> (Helgeson, 1994, p. 192)

The meeting is an important tool for managing the proposal process. Meetings can be held for such major tasks as conducting proposal strategy, making a bid/no bid decision, kicking off a proposal project, reviewing proposal storyboards or drafts, and analyzing cost estimates and strategy. Or they can be used for the more ordinary duties, such as proposal planning or status meetings. Regardless of their purpose, meetings are required to manage a proposal project. You cannot escape them, so you need to know some of the common traps to avoid.

Just counting the time of the people involved (and time is money in the business world), meetings can be very expensive to plan and conduct, yet they're often handled in a wasteful and unprofessional manner. Here are a few common problems:

- Meetings start late, often because key members arrive late.
- No specific agenda is sent out before the meeting, or the meeting has a vague purpose.
- When there is an agenda, it is often unrealistically long or too general.
- Some participants really don't belong there.
- Conversations aren't productive; they ramble on, often without resolution.
- The leader loses control, and the participants lose interest.
- Participants don't know how to reach for consensus; some members dominate, while others don't contribute at all.

Considering what's at stake, you have an obligation to make meetings work. As a leader, prepare carefully and provide firm direction. As a participant, contribute your expertise when it's needed and help the leader work toward consensus. Ten rules will help create meetings that accomplish their purpose. These rules fall into three stages:

- Stage #1: Before the Meeting (Rules #1–4)
- Stage #2: During the Meeting (Rules #5–9)
- Stage #3: After the Meeting (Rule #10)

Meeting Rule #1: Involve only necessary people.

"Necessary people" refers to only those who, because of their position or knowledge, can contribute significantly to the meeting. The size of your meeting group will, of course, vary according to the purpose of the meeting, but try to limit the invitation list. For those who don't have anything to contribute to the meeting but need to know what occurs in it, send them a copy of the minutes rather than inviting them to attend.

Meeting Rule #2: Send out an agenda.

An agenda gives people the chance to do some of their thinking beforehand and come prepared. As a leader, you can use the agenda during the meeting to avoid or limit digressions from the issues at hand. You may also want to list suggested time limits next to agenda items to keep discussions on track. On the agenda, clearly identify those people who will be expected at the meeting to provide information or report on the status of pending assignments.

Meeting Rule #3: Distribute reading material beforehand.

If you want participants to review reading materials in preparation for a meeting discussion, distribute them before the meeting. If you hand out the materials at the meeting, you can expect wasted time as the participants do the homework that should have been done before the meeting began. If the success of the meeting depends on an understanding of this material, help the meeting participants prepare for the meeting.

Meeting Rule #4: Have only one meeting leader.

Someone must be in charge at every meeting. Leaderless meetings are like rudderless ships—they almost always founder. Quite frankly, not everyone is able to run meetings effectively. Good meeting leaders must be able to

- Listen carefully, so that all views get a fair hearing
- Paraphrase accurately, so that earlier points can be brought back into the discussion when appropriate
- Give credit often, so that participants receive reinforcement for their helpful contributions
- Build on ideas, so that diverse points made by different participants work together to form a consensus

Meeting Rule #5: Start and end on time.

Once you show participants that meetings will begin promptly, latecomers will mend their ways. Colleagues who are notoriously late for their meetings have no incentive to be prompt if everyone waits for them. Allowing this practice to continue irritates those who arrive on time.

Setting a completion time beforehand makes it clear that discussions will have a limit. Most participants lose interest and energy after an hour. Routine meetings should stay under 90 minutes. For longer planning sessions, include hourly breaks and specific agendas with clear goals.

Meeting Rule #6: Keep meetings on track.

Open, lively discussions bring ideas to the fore. As a leader, however, you may need to take control when discussions move off the point. Be assertive—yet tactful—in discouraging these time-wasters:

- Long-winded digressions from the main purpose(s) of the meeting
- Discussion of topics or actions that involve few of the meeting participants
- Domination by a few participants
- Interruptions from outside the meeting, such as phone calls or pages

By following the agenda, the leader can redirect the focus of the meeting back to the task at hand. If a new issue is raised that deserves more attention, suggest that it be discussed further after the meeting or that it be an agenda item for another meeting. If participants must leave the meeting early, consider an adjustment to the meeting agenda to accommodate their early departure.

Strong leadership gets results, and results gain you the respect of participants who also want to accomplish the meeting objectives while avoiding inefficient use of their valuable time.

Meeting Rule #7: Strive for consensus.

When possible, reach decisions that all members can accept. A compromise is preferable to a group decision forced by a vote on alternatives. For the most productivity and follow-up support, meeting participants should feel they have contributed to meeting decisions, rather than having the decisions forced on them.

Meeting Rule #8: Use visuals.

As with oral presentations, graphics can help crystallize points for the participants. They are particularly useful for recording ideas that are being generated rapidly during the discussion. You may want someone outside the discussion to write these points on a flip chart, chalkboard, or overhead transparency. A laptop computer can also be used to write meeting notes that are projected on a screen with an LCD projector.

Meeting Rule #9: End with a summary.

Before the meeting ends, summarize the discussion and what action will be taken, including who is responsible for that action. A summary helps participants get a perspective on the entire meeting and gives them the opportunity to qualify points with which they may differ. It also helps you avoid reaching a decision that people agree with but nobody feels responsible for implementing.

Meeting Rule #10: Distribute minutes soon after the meeting.

Write and distribute minutes within 48 hours after the meeting. If the meeting was worth an hour of the participants' time, it is important enough to record. If discussion items were controversial, have all committee members approve preliminary minutes before final distribution.

SUMMARY

Proposal development is a process, one that should be clearly defined, documented, and repeatable in your organization. The process can be supported by many functions and different ways to staff and organize your proposal teams. The following are possible organizational approaches to managing and producing a proposal:

1. Dedicated
2. Collateral
3. Project
4. Outsource
5. Hybrid

A documented proposal process, training for that process, and a schedule to implement the process are key elements of proposal preparation. Proposal training can be provided by internal and external sources. The proposal directive contains the written proposal procedures and plans to guide the proposal team. In addition to helping the proposal team plan its activity, the schedule allows the team to monitor and control the progress of its tasks. Use the following tips to develop effective schedules for preparing your proposals:

1. Wisely use the time allowed for proposal preparation.
2. Use task- and responsibility-based schedules driven by your proposal management plans.
3. Schedule a seven-day workweek only when it's necessary.
4. Avoid short-changing scheduled tasks performed early and late in the proposal process.
5. Be sensitive to what day of the week you schedule a milestone.
6. Look for opportunities to divide major task milestones into staggered, smaller tasks.
7. Monitor schedule progress, maintaining a balance of holding firm to a schedule and changing it when needed.

The proposal scheduling and process should reflect sound proposal management planning. Development and implement the follow proposal management plans:

1. Capture plan
2. Customer interface plan
3. Facility and material plan
4. Staff plan

5. Communication plan
6. Document process plan
7. Quality assurance plan
8. Production plan
9. Security plan
10. Assembly and delivery plan

Meetings are important interpersonal links in the proposal process. Team-written proposals, in particular, require many meetings during the planning and writing stages. Because many people and much time are involved, ineffective meetings can waste time and money. Follow these rules for planning and running a meeting, and for doing follow-up work after the meeting:

1. Involve only necessary people.
2. Send out an agenda.
3. Distribute reading material beforehand.
4. Have only one meeting leader.
5. Start and end on time.
6. Keep meetings on track.
7. Strive for consensus.
8. Use visuals.
9. End with a summary.
10. Distribute minutes soon after the meeting.

EXERCISES

The first three exercises can serve as the basis for a group discussion, an individual oral report, or an individual written report.

1. Proposal Organizations, Functions, and Planning: Simulated content.

Give your opinion on how the following factors can affect how a company prepares its sales proposals. Specifically describe how the factors can affect the organization and functions of its proposal staff and the features of its proposal management plans.

- Whether the company sells to commercial or U.S. government customers
- Size of the company's typical proposal and the average number of proposals the company produces in a year
- Employment level and facility size of the company, including whether it has field or satellite offices
- Type of product or service it sells, including how generic or tailored the product or service is for its customers
- How much the company works with subcontractors and vendors to provide products or services to its customers

2. Proposal Organization: Real content.

For a company in which you have been or are currently employed, which type of proposal organization described in this chapter would be most appropriate for that company's sales proposals? Explain why. If possible, describe the organization and processes actually used to produce the company's sales proposals.

3. Proposal Scheduling and Planning: Simulated content.

Using the scheduling tips and proposal management steps in this chapter, identify and schedule the milestones you'd recommend for the following types of proposal activity. In addition, describe how the assumed type of proposal organization would influence your schedule.

- An unsolicited 30-page, one-volume sales proposal with a two-week development schedule from proposal start to submittal; assume the use of a collateral proposal organization
- A solicited 100-page, one-volume sales proposal with a 30-day development schedule from RFP release to proposal submittal; assume there will be no pre-RFP proposal development work and that a project proposal organization will be used
- A solicited, three-volume sales proposal (including a technical, management, and cost volume, each with 100 pages), with a 45-day pre-RFP preparation schedule followed by a 60-day proposal development schedule from formal RFP release to proposal submittal. During the first 15 days of the 45-day pre-RFP period, you will submit questions and comments to the customer about the draft RFP and by the end of that 45-day period you will complete a formal strategy plan, proposal outline, and proposal storyboards based on the draft RFP. During the 60-day period after formal RFP release, you will update your strategy, outline, and storyboard planning as required by the formal RFP and produce as many drafts of the proposal as you prefer; assume the use of a hybrid proposal organization, which will include several representatives from subcontractor companies and limited support of strategy and review consultants

4. Analyzing Meetings as a Participant: Real content.

In a class discussion, describe the good and bad features you have observed in business and social meetings in which you have participated. What could have been done to avoid the bad practices?

5. Analyzing a Meeting as an Observer: Real content.

Ask permission to sit in on a meeting in which you will not participate. Take notes on how the meeting does or does not follow the guidelines in this chapter; then write a report on your findings.

3 Preparing to Write

OBJECTIVES

- **Learn the importance of planning before you begin to write**
- **Learn to evaluate solicitations and customer requirements in preparing the proposal outline**
- **Learn to use a requirements matrix to ensure responsiveness to customer needs**
- **Learn to write effective outlines**
- **Learn to plan your proposal with storyboards and mock-ups**
- **Perform exercises using chapter guidelines**

Writing a successful proposal is a process. Good writers know that their rigor in completing this series of steps determines the quality of the final product, whether it's a proposal, letter, memorandum, report, or other document. You should spend about equal amounts of time on these three main steps in the writing process:

1. Planning your approach
2. Writing drafts
3. Reviewing and editing drafts

> The solution to most writing problems . . . lies not in laboring over words and phrases at the draft stage, or in searching for the magic words (which probably don't exist), but in spending more time in the stages of the writing process which come before and after writing the rough draft: organizing the information . . . and revising the rough draft.
>
> (Woolston, Robinson, and Kutzbach, 1988, reprinted 1990, p. 51)

This chapter describes the first step, the planning process. Chapter 4 describes the writing step, while Chapters 9 and 10 describe the review and editing steps, respectively.

In this chapter, we'll describe how to plan your proposal to meet content and organization requirements for a solicited or unsolicited proposal. For a solicited proposal in response to an RFP, expect the RFP to specify which topics are to be addressed in the proposal and how they are to be organized. For an unsolicited proposal, you may have little or no customer guidance about how the proposal should be written.

A proposal that doesn't meet customer requirements for content and organization is called nonresponsive and will probably fail. Such a proposal fails to give the details customers need to make an informed "buy" decision. Even if the proposal complies with all content requirements, a disorganized proposal—or one that doesn't follow customer instruction—can make it difficult to find information and can frustrate the reader into assuming the required topic isn't covered.

There are three major planning steps that will help plan the writing of your proposal:

1. Develop a summary outline based on customer guidelines for proposal content and organization, and develop a requirements matrix that contains the summary proposal outline cross-referenced to the applicable RFP and strategy plan requirements.
2. Using the requirements matrix, amplify the summary outline to further define the content of the various proposal sections.
3. Incorporate the amplified outline into a storyboard format to further refine the outline and integrate the planning of proposal text and graphics. As required, further amplify the storyboard outline and graphic ideas with a mock-up of the proposal layout.

Figure 3–1 summarizes the flow of these planning steps. The following sections describe these steps in greater detail.

THE SUMMARY OUTLINE AND REQUIREMENTS MATRIX

> A proposal must deliver critical ideas quickly and easily. Your writing must be clear if you want others to understand your project and become excited by it. It will be hard to accomplish this if you have not clarified your thoughts in advance.
>
> (Geever and McNeill, 1993, p. 17)

If your customer dictates the organization and content of the proposal, develop a summary outline to meet these guidelines. For a solicited proposal, these instructions are usually in the RFP. Follow these instructions even though you

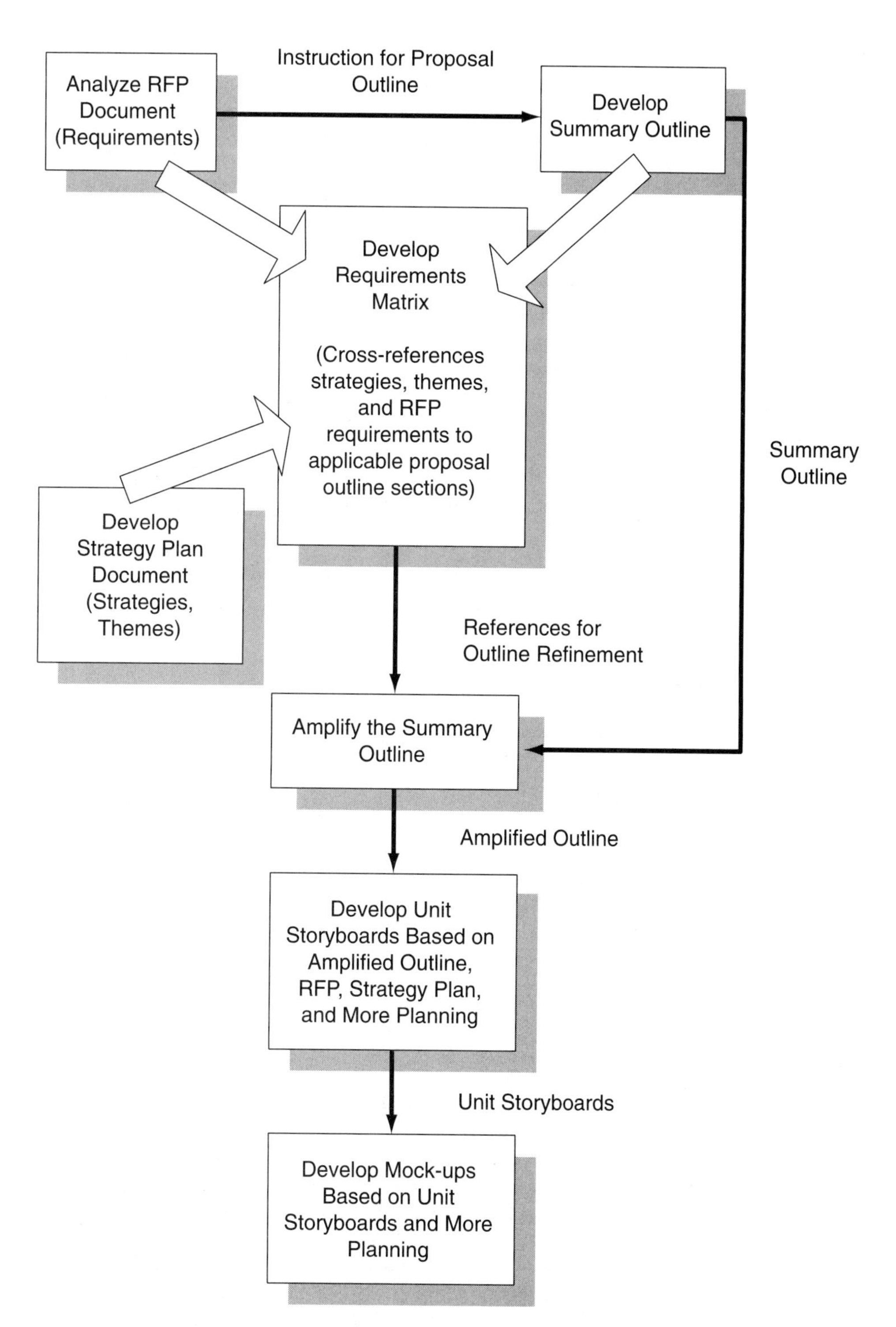

FIGURE 3–1
Planning steps before the writing begins

think that the approach won't produce a logical flow of ideas or allow you to rhetorically tell your story in the best way.

Some RFPs dictate broad categories of topics to be presented in a set sequence, while others provide very detailed directions for topic headings, subordination, and content. Still others call for proposal content focused on answers to a series of questions. Draft your summary outline based on the headings and topics requested in these instructions.

Follow RFP instructions to the point of using the names and spelling of section titles and the method of subordinating sections with Arabic or Roman numerals or alphanumerics. If you don't comply, the customer may find it difficult to find specific topics, and as you "customize" the order of proposal topics, you may inadvertently omit requested information. So resist the temptation to change or ignore these instructions, even if you think you're doing it for the customer's own good.

> Requests for proposals, RFPs, normally give detailed instructions concerning what elements to include in a proposal. You should follow all instructions exactly. A proposal submitted in response to an RFP is no place for "poetic license." Give them what they request or run the risk of having your proposal returned to you as incomplete or rejected out-of-hand as nonresponsive. The elements required in a proposal may vary somewhat from one RFP to another. This fact makes the following rule as absolute as gravity: *Read the instructions.*
>
> (Meador, 1991, pp. 19–20)

If you're planning an unsolicited proposal, ask the customer if there is a preferred approach. If there isn't, use your best judgment based on your needs and those of your customer. Chapters 7 and 8 have guidelines for organizing and writing proposals without customer proposal instructions.

The summary outline is amplified during the next planning step. However, before you amplify it, carefully analyze all sections of the RFP and your strategy plan to identify which sections of these documents pertain to the specific outline headings. To make this correlation, develop a requirements matrix, an example of which is shown in Figure 3–2.

The requirements matrix helps you to further plan proposal content as you amplify the summary outline and remain responsive to *all* RFP requirements. As shown in the Figure 3–2 example, the left column has the headings of the summary outline, while the right column lists all RFP sections that apply to the outline heading. If you have produced a strategy plan as described in Chapter 1, also use the matrix to show where to address strategies and themes in the proposal outline. The right column in the example matrix also lists applicable strategy plan sections.

Your proposal may not be a winner because it's offering the wrong product or service at the wrong price, but you don't want to lose because you haven't followed RFP instructions or addressed all RFP requirements. The requirements matrix will help you avoid these inexcusable mistakes.

REQUIREMENTS MATRIX	
Proposal Outline Heading	**Requirement Source and Summary**
1. Management Process	• RFP Sect. L.1 (Management Process) • RFP Sect. M.1 (Management Process) • RFP Sect. C.1 (Project Control)
1.1 Project Organization	• RFP Sect. L.1.1 (Project Organization) • RFP Sect. M.1.1 (Project Organization) • RFP Sect. C.1.1–C.1.1.4 (coordination of contract and customer management, use of integrated product teams, subcontract management structure, flow down of requirements to subcontractors) • Strategy plan Sect. 1.3 (theme—our streamlined project management structure with direct line to upper corporate management)
1.2 Key Project Personnel	• RFP Sect. L.1.2 (Key Project Personnel) • RFP Sect. M.1.2 (Key Project Personnel) • RFP Sect. C.1.2–C.1.2.2 (labor categories and qualifications, employee recruitment practices, advanced notice of management changes) • Strategy plan Sect. 2.1 (strategy—our approach to resolving customer concern for labor retention and training)
1.3 Project Scheduling	• RFP Sect. L.1.3 (Project Scheduling) • RFP Sect. M.1.3 • RFP Sect. C.1.3–C.1.3.3 (key project delivery dates, scheduling software, procedures for revising schedule) • Strategy plan Sect. 1.5 (theme—our aggressive schedule with preliminary planning already completed)
1.4 Project Budgeting and Accounting	• RFP Sect. L.1.4 (Project Budgeting and Accounting) • RFP Sect. M.1.4 (Project Budgeting and Accounting) • RFP Sect. C.1.4–C.1.4.5 (budgetary control of subcontractors, project cost control, procedures for projected cost overruns, accounting software, setting budgetary reserves) • Sect. H.1.2 (Cost Accounting Standards and Procedures) • Strategy plan Sect. 5.3 (competition strategy—our ghosting of competitor's prior bankruptcy and cost overruns)
1.5 Project Communication	• RFP Sect. L.1.5 (Project Communication) • RFP Sect. M.1.5 (Project Communication) • RFP Sect. C.1.5–C.1.5.4 (intranet and Internet software and architecture, videoconferencing resources, communication and data security) • Strategy plan Sect. 3.4 (concern strategy—our approach to resolving our concern for effective communication among all project team subcontractors and vendors)
1.6 Project Status Reporting	• RFP Sect. L.1.6 (Project Status Reporting) • RFP Sect. M.1.6 (Project Status Reporting) • RFP Sect. C.1.6–C.1.6.4 (procedures, content, and distribution of cost, schedule, and performance reports) • Strategy plan Sect. 1.4 (theme—our on-line, customized report generator to provide real-time project status)

FIGURE 3–2
Requirements matrix example

THE AMPLIFIED OUTLINE

> If you think of an outline as the specifications for a report and a draft as a prototype, the storyboards are the blueprints. They are a means of showing *more detail than an outline,* while retaining a form that permits easy change.
>
> (Woolston, Robinson, and Kutzbach, 1988, reprinted 1990, p. 17)

When you finish the requirements matrix, now it's time to refine the summary outline. Using the RFP, strategy plan, and other analyses of your readers' needs, expand your summary outline with more detail. Within the structure of your summary outline, use the following steps to produce an amplified outline for a responsive and organized proposal. (These steps can also be used as a starting point for proposals that have no customer guidelines for proposal organization or content.)

■ *Outline Step #1: Brainstorm on the subject of your proposal.*

Jot down all the ideas that enter your head about the proposal's content—including those from the RFP, strategy plan, and your understanding of your product or service. Focus on the problem your proposal addresses and the proposed solution. At this stage, don't concern yourself with wording, redundancy, sequence of ideas, or other fine points. Also, develop only the ideas for the body of the proposal. Save the beginning and ending sections until later; they are easier to outline and write after you have dealt with the main proposal.

■ *Outline Step #2: Distinguish major ideas from minor ones, deleting weak or redundant points.*

Use the following process for imposing order on a page of unorganized notes created during Outline Step #1:

- Highlight three to seven main points that will serve as basic building blocks of the proposal. (It's thought that readers best remember points grouped in three to seven units.)
- Connect subordinate points with arrows to the main points they support.
- Delete redundant or weak points.
- Number the highlighted main points according to their anticipated sequence in the proposal.

If your outline seems disjointed and chaotic, don't worry. This is typical for a developing outline. Keep working and you can wrest form from chaos.

■ *Outline Step #3: Rewrite the outline.*

To write your draft, you'll probably need a smoother outline than the one resulting from Outline Step #2. In your rewrite,

- Place points in the order they will appear in your draft.
- Cluster minor points below their respective major points.
- Show relationships and topical subordination with a numbering system (Arabic or Roman numeral), an alphanumeric system (A.1, A.2 etc.), or any other suitable arrangement. You can also define relationships and subordination with dashes, indention, or underlining.
- Put all points either in topic or sentence form.

Concerning the last point, topic form is preferable. Topic outlines comprise fragmented ideas rather than sentences—for example, "10 previous projects at site" as opposed to "Our firm has completed 10 previous projects at the site." Compared with sentence outlines, topic outlines are more concise, quicker to write, and less likely to lead to text that will have to be changed later.

■ *Outline Step #4: Check the outline for development and clarity.*

At this stage, evaluate how well you have developed the outline. Further expand on an idea if its support appears weak. Ensure that any subdivided point has at least two subordinate points.

■ *Outline Step #5: Change your outline as needed, but do so with care.*

An outline may change when you amplify it or begin the always creative and therefore unpredictable process of writing a draft. However, make outline changes with care. Once an outline is responsive to the RFP requirements, ensure that a revision doesn't delete a required topic or change its required location in the proposal. For a team-written proposal, outlines for individual writing assignments need to be integrated for consistent content, flow, and internal cross-referencing. Therefore, be certain that an outline change in one section doesn't adversely affect the content of another section.

STORYBOARDS

Whether a proposal is being written by one person or a team, it's recommended that the amplified outline lead to a technique called storyboarding, a process derived from a practice first used in Hollywood for writing movie scripts. The classic

> The storyboard is a writing worksheet that helps the authors organize their stories into main discussion points or topics. The storyboard enables the planning of each topic around the thesis-theme-visual elements of a short persuasive essay (i.e., the natural passage of technical discourse). The purpose is to help the authors discover their stories, help them be more pointed and purposive (i.e., encourage a problem-versus-solution slant in their arguments), and let the proposal team review the material before it is written out in a hard-to-follow and hard-to-change draft.
>
> (Tracey, 1993, p. 54)

storyboard process summarizes ideas in two-page storyboards. Figure 3–3 shows a simple example of a two-page storyboard, developed with the following five parts:

1. Information that identifies the proposal
2. A statement of the storyboard's topic
3. A summary sentence for the entire storyboard
4. Key sentences for developing paragraphs later in the draft (a form of an amplified outline)
5. A rough sketch of one complementary visual

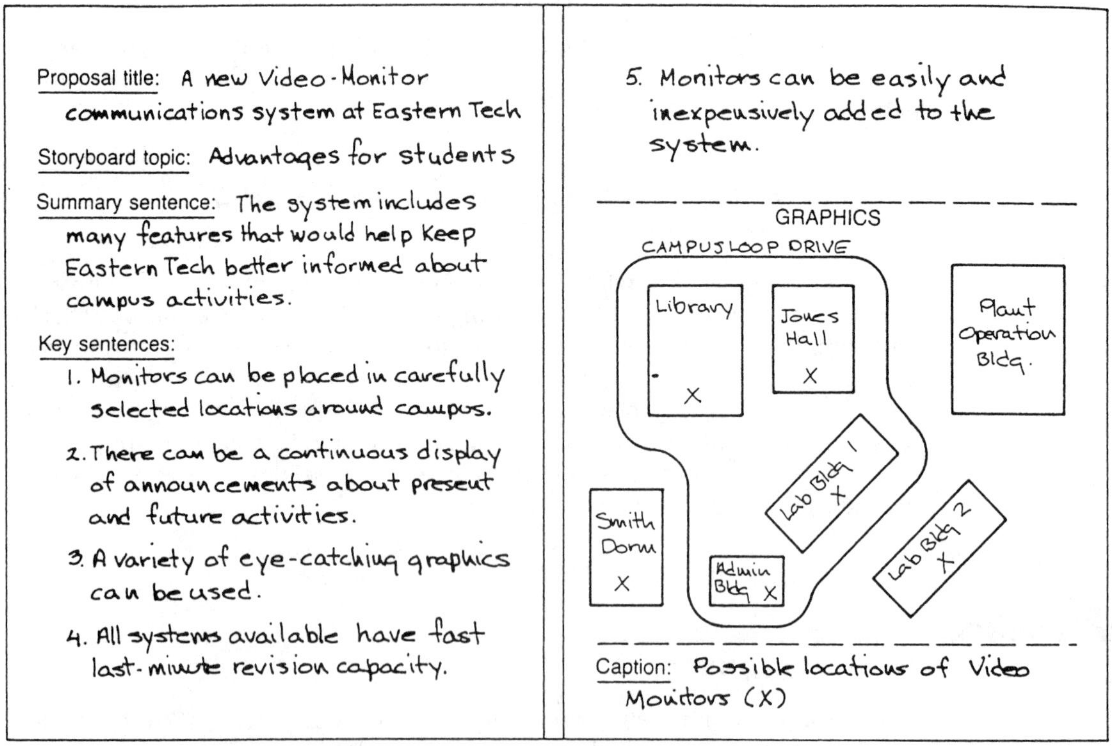

Proposal title: A new Video-Monitor communications system at Eastern Tech

Storyboard topic: Advantages for students

Summary sentence: The system includes many features that would help keep Eastern Tech better informed about campus activities.

Key sentences:

1. Monitors can be placed in carefully selected locations around campus.
2. There can be a continuous display of announcements about present and future activities.
3. A variety of eye-catching graphics can be used.
4. All systems available have fast last-minute revision capacity.
5. Monitors can be easily and inexpensively added to the system.

GRAPHICS

CAMPUS LOOP DRIVE

Library X

Jones Hall X

Plant Operation Bldg.

Lab Bldg 1 X

Lab Bldg 2 X

Smith Dorm X

Admin Bldg X

Caption: Possible locations of Video Monitors (X)

FIGURE 3–3
Two-page storyboard

Storyboarding involves a rigorous procedure. Although it requires the joint efforts of everyone on the team, it should be overseen by one person—for example, the proposal manager. It's also best to assign only one author per storyboard because this clearly denotes responsibility and avoids the problems of multiauthor coordination. Without these controls, the process can generate into the kind of time-wasting free-for-all that characterizes so many team-writing efforts.

In practice, many types of proposal storyboards are used. They vary in how many pages they represent in the proposal and what planning information they contain. For example, topic and page limit requirements in an RFP may not allow the division of topics into classic two-page storyboard units. Instead, one storyboard unit might need to represent five proposal pages to adequately address a complex topic. In contrast, a storyboard unit might represent less than two pages because of an imposed page limit or simply because two pages aren't needed to cover the topic. Additionally, a storyboard doesn't need to be limited to the five parts shown in the Figure 3–3 example. For example, Figure 3–4 shows an eight-part form. It reflects the contributions of many storyboard designs, including those produced by the following innovators in the proposal profession:

- Hughes Aircraft—the Sequential Topical Organization of Proposals (STOP) system
- Hy Silver—the Scenario Section Plan
- Shipley Associates—the Proposal Development Worksheet (PDW)

Regardless of their format, the common goal of storyboards is to plan the content of your text and graphics before the writing begins. Be creative; develop storyboard forms that best suit your needs. Forms can be available as printed worksheets or in computer template files.

After the summary outline and requirements matrix are completed, use the following steps to develop a storyboard based on the unit form in the Figure 3–4 example. (In the figure, the number of the storyboard step is labeled on the applicable part of the storyboard form. The following steps are general guidelines, targeted for planning a large, solicited sales proposal. However, the steps can be tailored for use with other types of proposals. You don't have to perform the steps in the order listed; follow a sequence that works best for you.)

■ *Storyboard Step #1: Provide information for proposal administration.*

Identify the proposal, volume, or unit title/heading; the name of the author and, as applicable, the volume leader; the page limitation of the unit; the date of the storyboard; and the date of approval. Show storyboard approval by initialing the signoff block.

■ *Storyboard Step #2: Identify applicable unit requirements.*

List all applicable requirements from the RFP, strategy plan, or other sources that drive the content of the storyboard unit. This list can include either a short reference

Unit Storyboard Form

Author's Name: #1	Proposal/Volume Title: #1

Unit Heading(s) and Section Number(s) #1

Date Submitted #1	Approval Signoff #1	Approval Date #1

Unit Requirements #2

RFP/Strategy Plan Section	Requirement Summary

Major Proposal Themes #3

Unit Themes #4

Unit Features #5	Unit Benefits #5

Unit Purpose and Summary #6

FIGURE 3–4
Eight-part storyboard form

Heading, Topical Outline, Graphic(s), Estimated Page Count: #7

Graphics #8

Designation, Size: Caption:	Designation, Size: Caption:
Description	Description

FIGURE 3–4
Continued

to the source, such as the section number of the RFP or strategy plan, or a written explanation of the requirement.

■ *Storyboard Step #3: Identify applicable major proposal themes.*

List the major themes (selling points) of the overall proposal that are applicable to the topic(s) in the storyboard unit. Use these major themes to develop the unit themes described in the next step. Major proposal themes should be identified in the strategy plan.

■ *Storyboard Step #4: Identify unit themes.*

List unit themes applicable to the topic(s) in your storyboard unit. When possible develop unit themes to support the major proposal themes. Unit themes can also be identified in the strategy plan.

■ *Storyboard Step #5: Develop a unit features and benefits table.*

In a table format, list features and associated benefits of the product or service addressed in the storyboard unit. This step helps to focus on the selling of the product or service, and provides a graphic that can be used to introduce and summarize the unit topic.

■ *Storyboard Step #6: Draft a short unit summary.*

In a short paragraph, summarize the content and major thrust of the unit. Much like the thesis sentence or paragraph for an essay, it gives you direction for developing the text and graphic(s). Don't spend too much time trying to get the perfect section summary. View it mainly as a planning tool for your writing, not as a finished written product. However, if written well, it can be used as an introduction in your proposal text.

■ *Storyboard Step #7: Insert and refine the amplified outline.*

Insert your amplified outline, indicating where the planned graphic(s) will be used. As required, revise the amplified outline, maintaining its topic outline structure. Estimate the page allocation for outline heading/section, including the space required for the graphics. The amplified outline should reflect the requirements from Storyboard Step #2, and the themes from Storyboard Steps #3 and #4.

■ *Storyboard Step #8: Show the conceptual plan for unit graphics.*

Based on the amplified outline, identify planned graphic(s) for the storyboard unit. The objective is to conceptualize the graphic. Although you can show a completed graphic, it's acceptable to provide a written summary of what the graphic will contain, a rough sketch of the graphic, or a copy of an existing graphic that is marked to show how it will be modified. Estimate the page size of each graphic (estimates of 1/4, 1/2, full, or foldout are normally sufficient) and assign a graphic identifier and caption. Chapter 5 contains guidelines for making proposal graphics, identifying and tracking graphics, and developing action captions. Ensure that each unit storyboard has at least one graphic.

■ *Storyboard Step #9: Review and approve the storyboard.*

When all the storyboards are drafted, assign a team or person to review and approve them before the authors proceed to the next proposal phase. Reviews can be formal or informal.

For larger proposals, it's best to have a formal review in which a dedicated team reviews and approves the storyboard units. In this type of review, the authors can be required to make an oral presentation about the storyboard unit as its image is projected on a conference room screen. For a smaller proposal with fewer authors, it might be just as effective to have the proposal manager work one-on-one with the author to conduct the review and grant approval. It can also be helpful for the storyboard units to be displayed sequentially in a dedicated proposal area to allow the entire proposal team to review and comment on the storyboards.

If the storyboard needs revision before it can be approved, instruct the author to develop another draft for review. When finally approved, it then becomes the basis for additional planning or writing the first draft.

The Benefits of Storyboards

Storyboarding can improve your proposals, but it's a process that requires a commitment and effort from the proposal management and writers.

The main benefits of storyboarding are that it integrates text and art in early proposal planning and leads to proposal sections that address all customer requirements and your proposal strategies and themes. Other benefits include the following:

- Team members find it easier to grasp and then comment on simple outlined points than to evaluate draft proposal text. The storyboard review also permits team authors to see how their units mesh with other parts of the proposal document.
- Outlines and graphic concepts take less time to develop than text and graphics. Quickly made and reviewed, they allow the identification and correction of content problems before the author begins producing text and graphics.
- Authors are also less likely to react defensively to criticism about their outlines than they are to their text because (1) they will probably have less "writer ego" involvement in their outlines, and (2) revising an outline will likely take them less time to do than a corresponding change to a written draft with finished text and graphics.
- Drafts undergo fewer revisions, because they flow from agreed-upon outlines, responsive to applicable RFP and strategy plan requirements.

The Challenges of Storyboards

There are also challenges in the storyboard process:

- The system takes planning and coordination. The authors must be trained to use the process before they are asked to follow it for an actual proposal. If committed to the process, the team should take no storyboarding shortcuts, no matter how frenzied the proposal-writing situation.
- Some proposal authors may be forced to accept a culture change in their writing habits—especially those who prefer to develop their writings through numerous drafts without the use of outlines. For those who have problems outlining and conceptualizing graphics, storyboards can be a frustrating—and seemingly impossible—chore. This resistance can be difficult to overcome.
- The process requires strong proposal management with excellent writing and interpersonal skills, to coordinate a team of authors and to overcome typical resistance to using storyboards.

If you're willing to face these challenges, storyboarding can help you create proposals that win.

MOCK-UPS

Time permitting, you can continue prewriting planning by developing mock-ups of the proposal content planned in the storyboard units. This additional step allows you to plan the appearance of the text and graphics on the finished proposal page. A mock-up example is shown in Figure 3–5. It displays a mock-up that might be used to plan page 6 of the model formal proposal in Chapter 7 (textbook page 206).

The mock-up allows you to further refine the details of your outline and graphics, as well as to plan the unit page allocation. On the mock-up layout, more topic headings and bulleted phrases can be added to the storyboard outline. The mock-up can also show a more detailed version of the planned graphics. However, the intent is to continue unit planning, not to display finished text and graphics. To show page allocation, the mock-up shows how much of the page will be used for the text or graphic. In the Figure 3–5 mock-up, note that arrows and outline details are used to estimate where text will appear on the page, while graphic borders mark the expected placement and size of the graphic. For proposal units that have a page allowance, mock-ups help you judge if the planned content is within the page limit.

Seasled Proposal, Author: Steve Wilson, Page 6

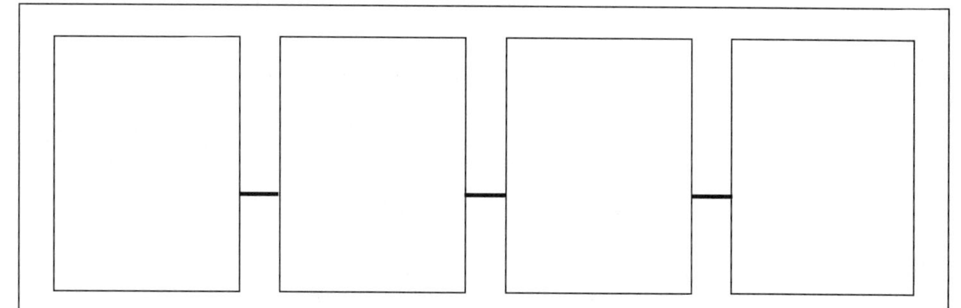

Flow diagram to show the process for:

- lowering the vehicles in the water
- boarding and positioning the vehicles by the divers
- cleaning the hull, including standby diver assistance and shipboard support

Figure 2. Hull Cleaning Process

2.4 Diver Rotation for 24-Hour Coverage
- A description of the 24-hour schedule for the divers, including: standby, on-duty, off-duty, and on-call periods

- Rotation schedule: promotes safety and productivity

2.5 Completion Times for Responsive Service
- Referral to Table 1 to show typical time for cleaning the galaxy-class tanker hull
- Average time: 80 hours

FIGURE 3–5
Mock-up example

SUMMARY

There are three major steps that will provide you with valuable planning tools for your writing:

1. Develop a summary outline and requirements matrix.
2. Develop an amplified outline. Use the following steps to develop an amplified outline based on the structure of the summary outline:
   - Brainstorm on the subject of your proposal.
   - Distinguish major ideas from minor ones, deleting weak or redundant points.
   - Rewrite the outline.
   - Check the outline for development and clarity.
   - Change your outline as needed, but do so with care.
3. Develop storyboards and mock-ups. Use the following steps to develop a storyboard:
   - Provide information for proposal administration.
   - Identify applicable unit requirements.
   - Identify applicable major proposal themes.
   - Identify unit themes.
   - Develop a unit features and benefits table.
   - Draft a short unit summary.
   - Insert and refine the amplified outline.
   - Show the conceptual plan for unit graphics.
   - Review and approve the storyboard.

EXERCISES

1. Preparing an Outline: Simulation

Assume that you are one of 20 project managers for a 500-employee firm that builds small shopping centers in the Midwest. Besides the main office in St. Louis, where you work, there are smaller offices in St. Paul, Chicago, Dayton, Fort Wayne, Topeka, and Dubuque. Until now, the company has survived without an in-house safety engineer. Increased business and liability, however, have convinced you to propose hiring a full-time safety engineer.

You have determined that your main proposal readers will be the vice president for engineering, vice president for finance, and personnel director. After analyzing their needs, you complete Outline Step #1—that is, you brainstorm and write down all the major and minor points your proposal should include, in no particular order. (The result is the following jumbled list.) Using this list as your starting point, complete Outline Step #2 (distinguish major ideas from minor ones, deleting weak or redundant points). Then rewrite the outline by (1) showing some points as subpoints of others, (2) putting ideas in their proper sequence, and (3) placing all points in topic form. In this case, place what you consider the most important information first in the rewritten outline.

- Cited by Occupational Health and Safety Administration (OSHA) for 10 safety violations last year
- One three-day job shutdown by OSHA
- Workers often don't wear hard hats or other safety gear.
- Safety training has been infrequent.
- Most new employees learn what little they do know on the job, not from training.
- Uneven quality in the courses given by the few outside consultants that have been used.
- Exit interviews with employees leaving the firm indicate on-the-job safety is one of the main reasons for leaving.
- Equipment insurance costs escalated 15 percent last year—agent says that two-thirds of that increase was due to excessive accidents.
- Safety engineer can do training at all offices.
- Morale will improve at all levels when employees see the firm's safety commitment.
- Hazardous wastes have been found on some construction sites—no one knows how to detect or handle them.
- Some large clients have expressed surprise that the company doesn't have a safety specialist.
- Joe Nunn, the engineer with the best safety background, is spread thin trying to act as a safety engineer (which he isn't) and doing his main job—supervising construction projects.
- Office workers could benefit from preventive programs on smoking, alcohol, drug abuse, etc.
- Need someone who can talk the same language as the insurance agents and OSHA
- Can hire good safety engineer for about $50,000 (about the same as construction engineers with a few years of experience)
- Safety engineers could train, observe projects, negotiate insurance rates, etc.

2. *Preparing a Storyboard: Simulation*

Based on the outline from Exercise #1 and the guidelines in this chapter, develop a storyboard to further plan your proposal to hire a full-time safety engineer. Assume that your storyboard unit will plan the entire proposal which has a five-page limit and that the storyboard will include the following parts:

- Proposal themes—the key selling points of the proposal
- Amplified outline—a more detailed topic outline of the summary outline developed for Exercise #1, including graphic ideas and page estimates for the text and graphics
- Graphic plans—at least two planned graphics with their size estimate and captions
- Features and benefits—at least four sets of features and associated benefits of your proposal, listed in a table format
- Proposal summary—a one-paragraph text summary of the proposal content and purpose

4 Writing

OBJECTIVES

- **Learn to analyze and better understand your readers' needs**
- **Learn organization principles for writing**
- **Learn basic rules for effective writing**
- **Learn specific principles for effective proposal writing**
- **Learn to overcome writer's block**
- **Perform exercises using chapter guidelines**

Good proposal writing is more than the creation of readable, concise, and grammatically correct text. It also relies on how well you understand and respond to reader needs and how easily the reader can find and understand key points in your text. Therefore, this chapter describes how to analyze the needs of your audience and then how to organize and write text to respond clearly, completely, and persuasively to those needs. It also provides tips for overcoming a problem that most writers experience—writer's block. (Although this chapter can help you improve your writing skills, we believe *some* writing skills come from the gene pool, not from instructions in a textbook.)

UNDERSTANDING YOUR READER

> Developing a clear, comprehensive picture of what the client is seeking is the single most important part of your whole proposal preparation process—if you get the requirement wrong, you'll get the solution wrong.
>
> (McCann, 1995, p. 53)

As you plan to write your proposal, you must answer a very important question: exactly what do your readers want to read? To answer this question, we'll describe obstacles proposal readers face, strategies to analyze your reader, and techniques for incorporating such analysis into your proposal.

As stressed in Chapters 1 and 3, it's important to analyze customer needs as you develop the proposal strategy, outline, and storyboards—planning steps that *lead* to writing. Why then should we now focus on reader analysis in this chapter—one about writing? The reason is that it's during the writing process that you should refine and apply your knowledge of the reader.

Obstacles Readers Face

The following are major obstacles that proposal readers face and the effect these obstacles have on reader evaluation of your proposal:

- Impatience—Readers want to understand the key points of the proposal quickly. If information can't be found easily in a proposal because it's poorly organized or written, readers may quit looking for it and then give your proposal a poor evaluation.
- Time conflicts—Readers may have to balance reading your proposal with the performance of their regular work duties, thus making it difficult for them to conduct a thorough proposal review. This situation can be aggravated if your proposal is one of many competing proposals they must read or if they have to be secluded for long periods to conduct a complex evaluation.
- Interruptions—As a result of time conflict, readers often won't have time to read your proposal without interruption. Their concentration can be broken by the demands of their regular work, phone calls, meetings, and business (and personal) emergencies. If your proposal isn't written so that a reader can resume reading it easily after taking a break, it may not be read thoroughly.
- Lack of knowledge—Readers may lack knowledge about the proposal topic, making it difficult for them to understand and to maintain interest in your proposal. (Of course, some readers may be very familiar with the topic, challenging the writer to accommodate different levels of reader knowledge in the same proposal.)
- Shared decision-making—A writer also can be challenged to meet the needs of readers who have varying levels of influence on whether your proposal wins or loses. This award decision can be made by one or shared by several. Some readers may not be the decision makers, but pass on their comments and recommendations to those who are. Therefore, even if several key readers favor your proposal, they may have to convince others with different needs and interests to make your proposal a winner.

To overcome these obstacles, learn as much as possible about your expected readers before you begin writing. The next two sections will describe how this can be done.

Preproposal Communication with the Customer

For solicited or unsolicited proposals, accept offers from the customer to make preproposal presentations about your product or service. For solicited proposals, take advantage of customer offers to comment on draft RFPs or to respond to customer requests for information (RFIs). These activities can help you influence the final RFP requirements and gather valuable information for planning your proposal strategy. Plus, they can allow you to become personally known by customers and to learn about their personal likes and dislikes.

To communicate better with your customer, polish your listening skills. People like to discuss their problems—sometimes more than they want to know the solutions for those problems. Before they'll see you as the right person to help them, you must first show an interest in understanding their concerns. To show this interest, get to know your customers with the following techniques:

- Spend most of your time asking questions in early conversations, rather than talking about what you have to offer. Come prepared with questions, being careful not to be so "prepared" that you fail to listen to the answers or to follow up with questions that evolve during the conversation.
- Show patience in listening to responses, even when conversations depart from the immediate subject at hand. After all, you may discover additional needs that you can offer to provide in other proposals.
- Take careful notes, both to demonstrate the importance you place on their comments and to collect information you can use later to write the proposal. For each conversation with the customer, summarize your findings in a contact report and then distribute it to your proposal team. File contact reports for use by future proposal teams.
- Present formal or informal briefings that showcase ideas you have for the proposed product or service. Noting the response of your customer can reinforce your approach or lead you to change your strategy. By working early with customers, you also might get them to "buy in" to your idea even before it's offered in the proposal. Your briefing can also give you a competitive advantage if customers use your information to develop requirements in their future RFPs.

During preproposal discussions, customers may also drop hints about what they prefer for the content, format, and organization of the proposal. Thus, even before you begin to write the proposal, you may find that you already have the answers to such questions as the following:

- What features of the idea should be stressed?
- How important is the project schedule?
- Who should be involved in the work?
- How many product or service options should be presented?
- Is a formal or an informal proposal format appropriate?
- How many pages should the proposal have?

This information is especially important for writing an unsolicited proposal for which you won't have written preparation instructions provided by the customer. But it can also help you get a head start on what will be a solicited proposal, allowing you to begin a pre-RFP proposal draft even before the RFP, including its proposal preparation instructions, is released.

By doing your homework—that is, by really listening to those whose problems you wish to solve—you stand the best chance of writing a responsive and winning proposal.

Analyzing Reader Needs

Before you write even the first line of your proposal, collect all you know about the reader, including what you've learned through personal contact and the reading of draft or final RFPs. (Even after the writing begins, take advantage of other chances to further analyze your readers as you have more customer contact and collect new marketing intelligence.) The following steps will help you analyze the reader as you prepare to write.

■ *Reader Analysis Step #1: Write down what you first know about your readers.*

Don't trust your memory to retain all the important details about your readers. Force yourself to record answers to these questions:

- What do they want from the proposal?
- What actions do you want them to take?
- What role do they have in their organization?
- What features of their professional training could affect their reading of the proposal?
- What features of their personal backgrounds or lifestyles could affect their responses to the proposal?

Don't worry if you can't answer all of these questions At this point, it's just as important to find out what you don't know. Proceed with your analysis and take the following steps.

■ *Reader Analysis Step #2: Read proposals written to the same readers or customers; talk to colleagues who developed those proposals.*

For planning an external sales proposal, review proposals your company has already submitted to the same readers—or, if you don't know the readers, to the target customer organization. Talk with people in your company who were involved with the development of those proposals. In addition, review the RFPs that generated these proposals and any related RFPs that were received from the same customer but didn't result in a proposal from your company. Collect any information that might help you better understand the customer's needs and preferences. For in-house proposals, it can be easier to learn about potential readers, because you have a better chance of knowing and having frequent contact with them.

Reader Analysis Step #3: Pinpoint the decision maker(s) in your audience.

As you try to identify customer decision makers, remember that readers can be classified by the degree of influence they have on the success of your proposal:

- Decision makers (first-level audience)—They make their own assessment or work in conjunction with the assessments of others.
- Advisers (second-level audience)—They don't make the final decision themselves. Instead, they read and evaluate the proposal and pass their assessment to the decision makers.

Of course, pay special attention to your first-level audience, but don't forget the second-level readers. Although they won't make the final decision, they can greatly affect it.

Reader Analysis Step #4: Determine audience knowledge of the proposal topics.

Proposal writing often requires the translation of technical or complex topics into language that nontechnical or generalist people can understand. To do so, you must first judge the interest and knowledge your reader has for the topic. The following categories will help you classify the interest and knowledge of a reader:

- Managers—These readers manage people, set budgets, and make decisions of all kinds. Thus, management readers may not be familiar with technical points of your proposal. Even if technical or specialist professionals become managers, once into management, they may be removed from the hands-on details of their profession.
- Experts or specialists—This reader group includes anyone with a good understanding of your topic. They may be well educated—as with engineers and scientists—or have an education earned from hands-on experience. A maintenance supervisor with years of experience, but with no college training, could be considered a technical "expert."
- Operators—This group includes the readers who are operators—people who would use the product or service offered in your proposal.
- Generalists—These readers, also called "laypersons," may know the least about your topic or field. This is a catch-all group of readers who don't fit into one of the other three categories. Like managers, the generalist readers may have broad knowledge of a topic.

Reader Analysis Step #5: Conduct a strategy meeting.

With the information gathered in the preceding steps, conduct a strategy meeting as described in Chapter 1. In this meeting, discuss customer concerns and issues and develop strategies to meet these needs. Formally document these strategies in a strategy plan.

If you've completed a successful audience analysis at this point, you're now ready to begin your writing by applying the organization principles and writing rules described in the remainder of this chapter.

THE PRINCIPLES OF ORGANIZATION

A proposal can affect a reader not only by the words it uses but also by the organization, display, and location of those words—the basis of what we call the "principles of organization," which are summarized in Figure 4–1. The following rules will help you apply these principles.

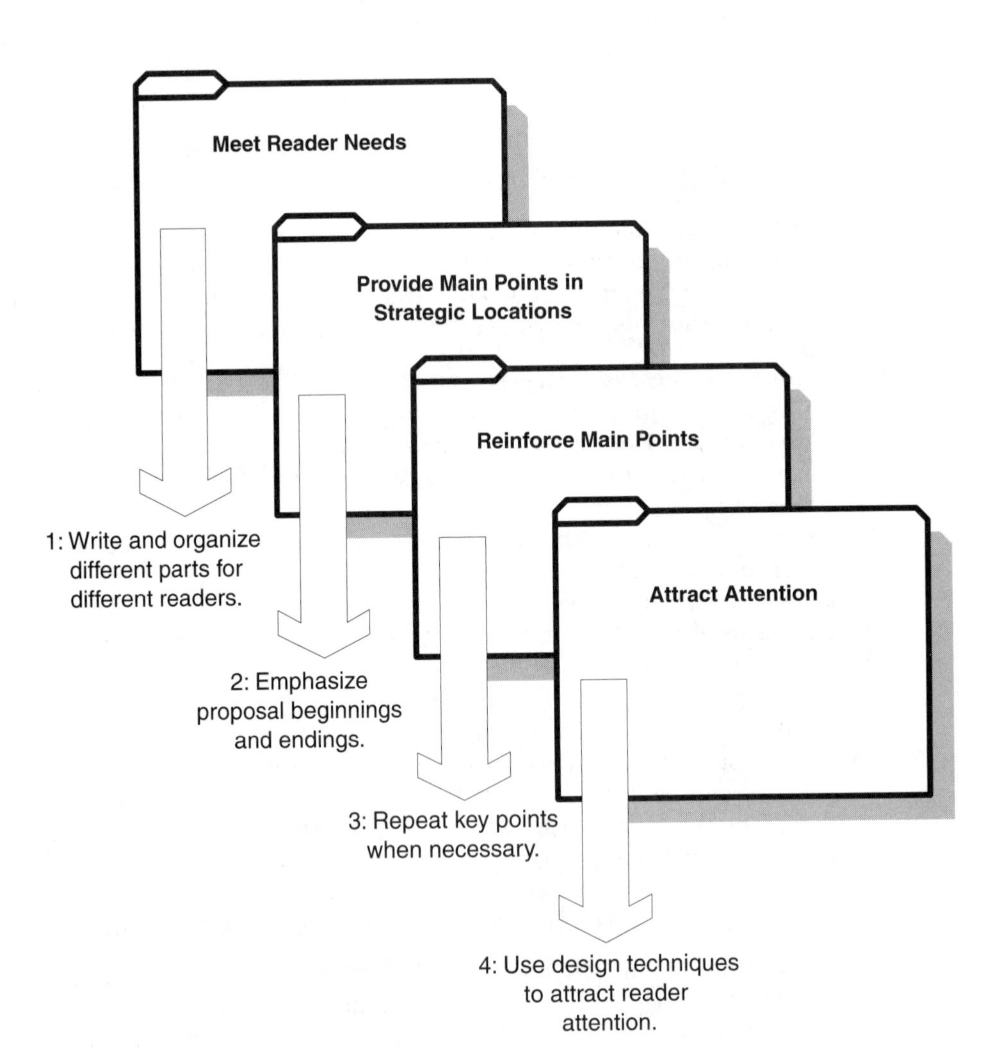

FIGURE 4–1
Principles of organization

■ *Organization Principle #1: Write and organize different parts for different readers.*

You can expect the reading of your proposal to be similar to the "channel surfing" done with a TV remote control. Channel surfers switch from channel to channel, depending on their interest and knowledge of the subject, the number of available channels, and their preferred method for scanning the channels.

Similarly, length and content of the proposal can affect what readers will read in the document. In addition, reader knowledge and reading patterns can affect which parts get read and in what order they are read. The longer and more complex the proposal, the less likely it is that a reader will read it from beginning to end. Indeed, in a proposal with many sections or several volumes, readers may be assigned to read only portions pertaining to their expertise.

As previously noted, readers can be identified by their knowledge. To help you target your readers effectively, Table 4–1 shows how reader needs can affect the content and organization of your proposal.

TABLE 4–1: Reader Categories and Needs

Category	Specific Needs in the Proposal
Managers	• Background information • Definitions of technical terms • Lists and other design devices that highlight points • A clear statement about what's supposed to happen next • Simple but informative summary graphics • A strong emphasis on responding to customer requirements and the benefits provided by the proposed product or service
Experts	• Thorough explanations of technical details • Detailed data placed in tables, figures, and schematics • References to outside sources, if used in the proposal • Clearly labeled attachments, appendices, and enclosures for supporting information
Operators	• An easy-to-use contents matrix for locating sections that relate to them • Easy-to-read listings for procedures or instructions • Definitions of technical terms and acronyms • A clear statement of how the proposed product or service will affect their job
Generalists	• Frequent use of graphics, such as charts and photographs • A clear distinction between facts and opinions • All the needs listed for managers

Regardless of their expertise, readers may "speed-read" your proposal. This reading method can take the following forms:

- Quick scan—Readers scan easy-to-read sections, such as executive summaries, introductory summaries, overviews, introductions, tables of contents, conclusions, and recommendations. They pay special attention to beginning and ending sections—especially in longer proposals—and to graphics.
- Focused search—Readers go directly to parts of the proposal that will meet their immediate needs for information. To find information quickly, they search for such format devices as subheadings and listings to guide their reading.
- Amplified search—Readers return to the document, when time permits, to read or reread important sections or to read attachments, appendices, or enclosures that provide more details about a topic in the main body of the proposal.

Therefore, your writing must accommodate readers who read sections based on their interest and knowledge and therefore don't necessarily read the sections in the order they appear. The basic guideline is this: organize and write each section of your proposal aimed at the audience most likely to read that section. This approach may lead to different levels of detail from section to section. As you vary the levels of detail, consider the use of stepping-off points for your readers, giving them the choice of reading on for more details or going to another part of the proposal.

Although the technical detail and stylistic features may change from section to section, your proposal must hang together as one piece of work. It might be written by a committee or for a committee, but it shouldn't suffer for that reason. Common threads of organization, theme, and tone must keep it from appearing fragmented or pieced together. The following organization principles will help you weave these threads.

Organization Principle #2: Emphasize proposal beginnings and endings.

> A frequent problem with business and technical writing is the tendency to lead *to*, rather than *from*, major ideas. Many writers believe that they have to build their case, that skeptical readers will not agree with their conclusions unless they first demonstrate how they arrived at those conclusions. This tendency results in documents that are unemphatic, difficult to follow, and filled with unnecessary detail.
>
> (Freeman and Bacon, 1990, reprinted 1995, p. 156)

Your proposal readers may be too busy to read the whole document, become bored with its details, or be interrupted while reading it. In fact, readers may not even want to read your proposal, but do so only because their boss assigned them the task.

In fiction books, writers rely on the interest and patience of readers, dropping hints through the narrative before finally revealing who did what to whom.

In contrast, your proposal readers probably won't read your proposal for its suspenseful entertainment value. Instead, expect proposal readers to want information in predictable locations without having to search for it.

Faced with this kind of proposal reader, do the following:

- Organize your proposal assuming it will be read carelessly.
- Write every portion of your proposal as if you're certain readers won't get beyond the first few sentences.
- Emphasize the beginnings and endings of proposal segments.

At the beginning of key proposal segments, let readers know the segment's main point and what "direction" the segment is leading them. To meet these expectations, use the *main* principle of organization—that is, place your main points in strategic parts of the proposal. Apply this principle with the following techniques:

- Begin the overall proposal with an overview section
- Begin each section with an overview paragraph or two
- Begin each paragraph with an overview (topic) sentence

Be in control; tell readers what you consider the most important point to be, rather than letting them guess on their own. By summarizing your point up front, you also alert the reader to the rhetorical flow or direction of your writing.

At the end of key proposal segments, don't let your writing simply drop off. Instead, give readers a conclusion or summary of what they've read or provide a transition to the next topic. Common transitions include linking or contrasting ideas in adjoining paragraphs or sections and briefly summarizing what information will be covered in the next portion of the proposal. When using this latter transition method, give your reader a sense of order and logic by consistently summarizing and presenting the information. Take, for example, an introduction in which you write that the following section will cover points A, B, C, and D. For a sense of order, ensure that these points are presented in the same order they were introduced.

This approach to beginnings and endings is similar to the preacher's maxim emphasized in Chapter 11 about speaking: "First you tell 'em what you're gonna tell 'em; then you tell 'em; and then you tell 'em what you told 'em." Of course, this maxim could be taken to extremes. If you do this for every paragraph, the proposal would be choppy and wordy. Summarize conclusions in the most strategic locations of your proposal, such as the ends of major sections.

■ *Organization Principle #3: Repeat key points when necessary.*

Assuming that few readers move straight through a proposal and that they often skip to sections that most interest them, you may need a redundant approach in your proposal. In other words, you may have to repeat information in your text and graphics to ensure that your selective readers don't miss an important point. Repetition can also help you reinforce a point after your readers see it the first time.

However, repetition should be done with care, for too much of it can distract and irritate readers. Excessive repetition can also make it difficult for you to keep your proposal within a page allowance. Will readers who read every word of your proposal be put off by the restatement of main points? Probably not, if you don't overdo it and if you strategically repeat information of interest, such as major features, benefits, recommendations, and conclusions.

Repetition can be helpful when writing a multivolume proposal. Consider, for example, the writing in one volume of a three-volume proposal. Because of the division of volume topics, you find you must refer readers to another volume for more details. In this case, it may be best to repeat the details because readers could be prohibited from seeing the referenced volume or simply not bother to read it, because they don't have ready access to it.

In contrast, it would be best to limit or even avoid repetition in a proposal that has few pages or must comply with a very restrictive page limit. In a short proposal, the repetition could easily appear excessive. In a page-limited proposal, repetition could take up space that would be better used with other information. In these two situations, instead of repeating information, direct readers to its original location in the proposal.

■ *Organization Principle #4: Use design techniques to attract reader attention.*

The display of your text can influence the content of your writing and your ability to attract readers' attention. The following design techniques will help you display your text to highlight and summarize key points in your proposal.

Text Lists. Use lists to describe or identify items or actions in a group or series. Almost any group of three or more related points can be made into a list. For example, a list can present

- Examples
- Reasons for a decision
- Conclusions
- Recommendations
- Steps in a process
- Key product features and benefits

Lists can help emphasize important points, because lists normally attract more attention than the text surrounding them. They also provide a change in format and allow you to cluster items for easier reading.

The most common visual clues for lists are numbers and bullets (enlarged dots or squares). Although there are many opinions about when to use numbers or bullets in lists, we offer the following rules, because they are simple and easy to apply:

- Use numbers when you need to indicate an ordering of steps, procedures, or ranked alternatives or priorities or when a number makes it easier to refer to that

item elsewhere in the proposal. (Be sure that the number is needed, because the reader may infer a reason for a number, even if that wasn't your intention.)
- Use bullets when there's no reason to number the items.

The best lists are those that subscribe to the rule of short-term memory. That is, people retain no more than five to nine items in their short-term memory. A listing of more than nine items may confuse rather than clarify an issue. Consider placing 10 or more items in two or three groupings to help the reader better grasp the information.

With lists it's possible to have too much of a good thing. Too many lists on one page can create a distracting, fragmented effect. A good rule of thumb is to use no more than one or two lists per page. Too many on the same page force readers to decide which one deserves their attention first.

Depending on purpose and substance, lists can be structured with words, phrases, or sentences. Whatever structure you choose, follow that approach in each listed item and pare down the wording as much as possible to retain the impact of the list format.

Text Emphasis. You may want to emphasize an important portion of text. Word processing programs typically give you the options for underlining, boldfacing, italicizing, and full capitalization. The least effective techniques are FULL CAPS because they can be difficult to read within a paragraph and can be distracting to the eye. The most effective are *italics,* underlining, and **boldface,** because they can add emphasis without distracting the reader. The italics and underlining methods can be especially effective because it's able to draw attention to text without the overemphasis of boldfacing, which can be hard to read.

Whatever technique you select, use it sparingly. Overuse can create a busy-looking page that leaves readers confused about what's really important to read. Excessive in-text emphasis also detracts from the impact of headings and subheadings, which should receive significant attention in your proposal.

Focus Boxes. Short sentences or lists bordered by a box can be placed in proposal margins or key breaking points in the proposal text, such as section headings. Similar to lists, these displays can attract readers who might be scanning a page and can help you emphasize key proposal points in the nearby text. These points can include applicable proposal themes or key features and benefits, or a summary of the adjoining text. To further differentiate it from the text, the focus box can be designed with the following features:

- Italicized or boldface text within the box
- Lists with bullets, numbers, check marks, squares, or other creative devices within the box
- A type size and font style within the box that are different from that used in the text
- Different sizes and shadow effects for the box borders

Chapter 5 provides more ideas about typographical and design features that can enhance proposal appearance.

BASIC WRITING RULES

Although this textbook isn't meant to be a writing primer, proposal writers should remember some basic writing techniques that are as applicable to proposal writing as they are to other types of writing. Strengthen your basic writing skills by using the following rules. The editing rules in Chapter 10 can also be used to improve your writing.

■ *Basic Writing Rule #1: Write with effective style.*

To augment your own writing style, use the following style rules:

- Be concise without sacrificing clarity; use short, concrete words and avoid clichés and wordy or trite phrases.
- Use strong, active-voice verbs for shorter and more forceful sentences and for a clearer understanding of the subject performing the action.
- Keep a customer perspective by the frequent use of words such as you and yours.
- Follow parallel form in lists and headings to avoid distracting changes in text structure.
- Vary the length and style of paragraphs and sentences for text variety, clarity, comprehension, and readability.

■ *Basic Writing Rule #2: Use correct grammar.*

To avoid common grammatical mistakes in proposal writing, observe the following rules:

- Make sentence subjects agree with verbs.
- Ensure that pronouns are clearly identified, agree in number with their antecedent noun, and accurately reflect the gender of the subject.
- Ensure that all sentence modifiers are clear, being especially careful to avoid dangling or misplaced modifiers.
- Properly use commas, understanding how comma placement can affect the meaning of a sentence.
- Avoid common errors in word usage.

■ *Basic Writing Rule #3: Write with clarity, conciseness, and accuracy.*

Write with clarity and conciseness. A common problem is writing a sentence that can be interpreted in more than one way. Although the writer knew the intended meaning when it was written, ambiguity forces the reader to select a meaning from among various choices—and the wrong one can be selected. Don't assume your readers have the knowledge you have or that they'll understand your intent because it was intuitively obvious to you. Try to imagine the reader peering over your shoulder as you write: your writing should be so lucid that the reader would understand it without asking for clarification.

Write accurately. When you draw material from a reference proposal or other document, use it only after you verify its accuracy. Never write anything that you know to be inaccurate (or just a little inaccurate) because you think readers won't read it or, if they do, won't question it. Also, if you write something that you don't understand, which can happen when you use information from other written sources, don't expect the reader to understand it.

■ *Basic Writing Rule #4: Use paragraphs as the framework of your writing.*

Paragraphs represent the basic building blocks of any document. Keep the typical length of paragraphs at around 6 to l0 lines. If you see that your topic requires more than 10 lines for its development, split the topic among two or more paragraphs. Shorter paragraphs can add white space to your page layout and help the readability of your proposal.

As summarized in Figure 4–2, most paragraphs contain the following elements:

- Topic sentence—This sentence states the main idea to be developed in the paragraph. Usually it appears first. Don't delay or bury the main point, for busy

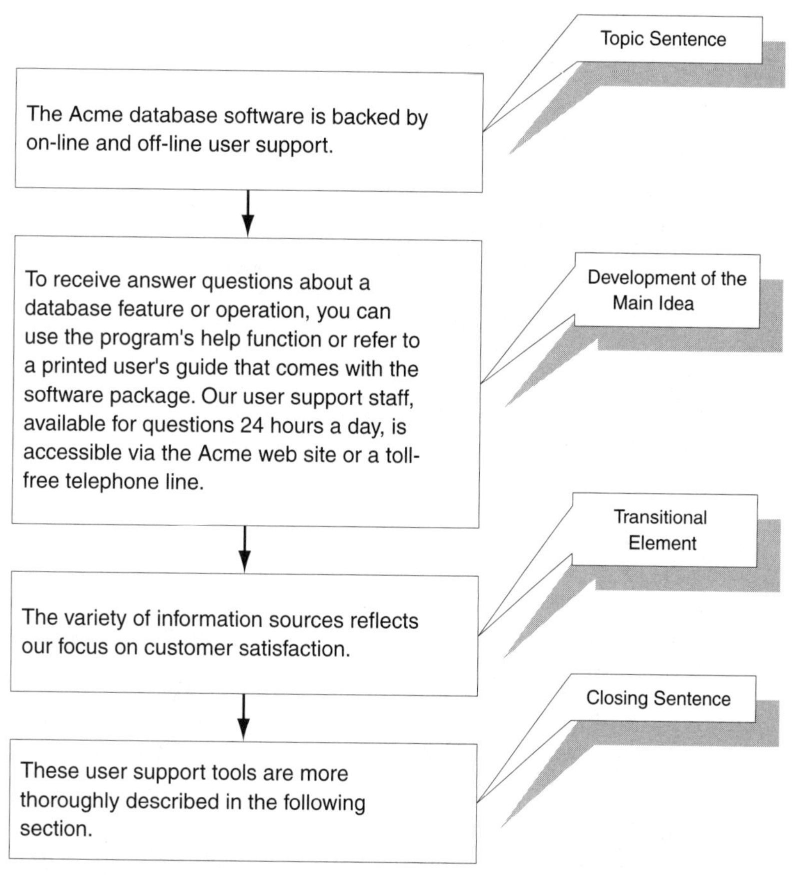

FIGURE 4–2
Elements of a paragraph

readers may read only the beginnings of paragraphs. If you fail to put the main point there, they may miss it entirely.
- Development of the main idea—Sentences that follow the topic sentence develop the main idea with examples, narrative, explanation, and other details. Give the reader concrete supporting details, not generalizations.
- Transitional elements—Transitions help the paragraph flow smoothly. Use transitions in the form of repeated nouns and pronouns, contrasting conjunctions, and introductory phrases. Transitions can also be sentences that continue the topical flow, summarize the main idea, or support a conclusion in the closing sentence.
- Closing sentence—Most paragraphs, like sections and documents, need closure. Use the last sentence for a concluding point about the topic or a transitional point that links the paragraph with the one following it.

However, there can be exceptions to these elements. For example, you may choose to delay the statement of a topic sentence until you engage the reader's attention with the first few sentences. In other cases, the paragraph may be short and serve only as an attention grabber or a transitional device between longer paragraphs.

PROPOSAL WRITING RULES

> If a proposal is to make a good first impression, the reader's first 10 seconds of exposure are crucial. The reader's first concerns are:
>
> - How long will this one take to evaluate?
> - Is it long and wordy or short and concise?
> - Is it well organized?
> - Who submitted the proposal?
>
> Content counts more than form, but first impressions are also created by a proposal's outward appearance, especially by evidence that it is well organized and easy to read.
>
> (Jacobs, Menker, and Shinaman, 1990, p. 124)

To further refine your writing skills, use the following rules to make your proposal writing responsive, informative, persuasive, and credible.

■ *Proposal Writing Rule #1: Follow your writing plan.*

Storyboards and mock-ups provide the writer a starting point and structure to follow during the writing process. Don't spend the time developing these valuable tools and then fail to follow their guidance as you write. These planning tools will help you

- Comply with RFP requirements and support your proposal strategies and themes
- Meet your target page allowance by avoiding the creation of text (and graphics) that can't be used because of space limitations
- Plan the content and integration of your text and graphics

Without the use of storyboards and mock-ups, writing can be marked with a series of drafts and rewrites that don't address RFP requirements or support your win capture strategy. As you write, if you believe the storyboard—especially its outline—should be revised, get approval from proposal management before making the change. Unilateral decisions by writers to change the approved outline can cause their writing to be nonresponsive to the RFP and to duplicate unnecessarily or contradict information that appears elsewhere in the proposal.

Proposal Writing Rule #2: Provide the basic proposal details.

Take a lesson from newspaper reporters who write to answer the following questions in their news articles: Who? What? When? Where? How? Why? Translated to proposal writing, the answers to these questions can provide the following basic details about the offered product or service:

- *Who* will provide it and *who* will use it?
- *What* does the customer want and *what* work is required to provide it?
- *When* will it be produced and delivered?
- *Where* will it be produced and delivered?
- *How* will it be produced and delivered, *how* will it benefit the customer, and *how* much will it cost?
- *Why* is it necessary and *why* are you taking this approach in providing it?

> Simply regurgitating the RFP buys you absolutely nothing; it will earn the evaluator's disgust with your ineptness, or worse yet, his everlasting hatred for your insult to his intelligence.
>
> (Helgeson, 1994, p. 99)

It can be argued that providing this detail can overcommit your company—especially if the proposal is accepted as a contract. The challenge is to provide enough detail to show you've done your homework without committing your company to something it can't provide. One way is to qualify details by noting that they are based on your preliminary planning and understanding of the customer's needs. For example, you could describe what you intend to do based on this planning and understanding and indicate that your final approach is subject to further discussion with the customer after contract award.

Others may balk at giving details—especially a company's proprietary data—because they fear losing the proposal competition and having this data used by the customer or passed to a competitor. There's nothing you can do to guarantee that this won't happen. Nevertheless, have your proposal clearly identify proprietary data and any limit on the use and distribution of this data. There can be situations in which proprietary detail should be left out because of a valid concern over its unauthorized use. However, a decision not to include proprietary information in a proposal could make your proposal nonresponsive and a loser if that information is needed to respond to an RFP requirement.

Be direct when explaining your proposed approach. For example, don't just claim it's your objective, philosophy, or policy to do something a certain way; say what you'll do, why you'll do it, and what benefit the customer will derive from it. (It may be your philosophy to do something, but that doesn't necessarily mean you will.) Also, don't simply state that something is important to do; explain why it is and how and when you'll do it. (Noting the importance of an action isn't the same as committing yourself to that action.)

Early text drafts often suffer from too much emphasis on *what* the customer wants, with few or sketchy answers for the other questions. This result is understandable because *what* the customer wants can be the highest concern (and rightfully so) of the proposal team early in its writing phase. Plus, what the customer wants can be obvious to a team that is responding to an RFP. What may not be as obvious are the answers to the other questions. These answers can take time to develop and be the difference between a winning and losing proposal.

■ *Proposal Writing Rule #3: Ensure your proposal details have purpose.*

Temper the depth of your proposal details by ensuring that everything you write has a purpose. Avoid the "data dumping" of unnecessary information that leads the reader to wonder, "Why is this in the proposal?"—known as the "So what?" test. Ensure that your proposal content passes this test.

The depth at which you provide the basic details described in Rule #2 can be affected by many factors, such as the following needs:

- To thoroughly respond to customer requests for specific information
- To write within a proposal page allowance dictated by your proposal team or the customer
- To provide information that meets the wants of readers possessing varied skills and interests
- To provide background information that helps readers better understand a point in your proposal
- To provide details that show you understand a topic or an issue in the hope of boosting your credibility with the readers

If you feel extra details are warranted in your proposal, but you want to give readers the choice of reading them, place those details in graphics, attachments, appendices, or enclosures. Don't scrimp on the details just because you think readers will already know them. Instead, provide details based on the following premises:

- Although the *readers* may know the details, it's important for you to gain readers' credibility by having them recognize that *you* also know the details.
- You get credit only for what appears in your proposal, not what you assume the customer knows without reading your proposal.

By following this rule and Proposal Writing Rules #1 and #2, it's more likely that your proposal will receive good *quantitative* and *qualitative* evaluations by the customer. In other words, not only will your proposal address all the topics and requirements it should (a quantitative goal), it will address them thoroughly and persuasively (a qualitative goal).

■ *Proposal Writing Rule #4: Use "boilerplate" information for your proposal, but use it carefully.*

As a starting point for a new proposal, it's common to use information from existing documentation. Generically called "boilerplate," it can include text and graphics from many documents, such as proposals, technical specifications, company procedures and directives, and advertising or marketing materials. Boilerplate that requires little modification or tailoring to be applicable to a new proposal can greatly reduce the time you spend on researching and writing for that proposal. An example would be a proposal description of a product or service that varies little among customers. However, even boilerplate that requires significant revision to be applicable can be a time-saver.

As helpful as it can be, boilerplate comes with risk. If the information isn't tailored enough for the subject customer, it can appear as "data dump" information that adds little or nothing to the proposal and "one size fits all" information that could have been written for *any* customer. Poorly tailored boilerplate wastes proposal space, reflects an unprofessional approach to proposal writing, and shows little concern for your readers' needs.

Without enough tailoring, boilerplate can fail to answer all the customer questions about your product or service. Reliance on boilerplate can also produce a proposal that conforms to the content and organization of the boilerplate source, rather than to the customer's proposal instructions for the subject proposal. In the extreme, the careless use of boilerplate can embarrassingly result in the delivery of a proposal that still has the name of the customer for which the source proposal was originally written.

What's a proposal writer to do? The answer is to use boilerplate but to adapt it to the target customer. As summarized in Figure 4–3, the following are ways to tailor your boilerplate:

- Comply with Proposal Writing Rule #2. By ensuring that your boilerplate is tailored to address the who, what, where, when, how, and why of the subject proposal, you should be able to provide the details your customers want to know about your proposed product or service.
- As a planning tool, use a requirements matrix to pinpoint clearly where all RFP requirements and all of your proposal strategies and themes are to be addressed in the proposal—including the boilerplate you use. Also follow Proposal Writing Rule #1 to ensure your proposal is responsive to the RFP and addresses your strategies and themes.
- Include a compliance matrix in the proposal to indicate to readers where each RFP requirement is covered in the proposal. This step shows your responsiveness to customer requirements and your concern for the readers' ability to find topics in the proposal. A compliance matrix also provides an audit check for you to ensure that the proposal has addressed all RFP requirements before the proposal is delivered.

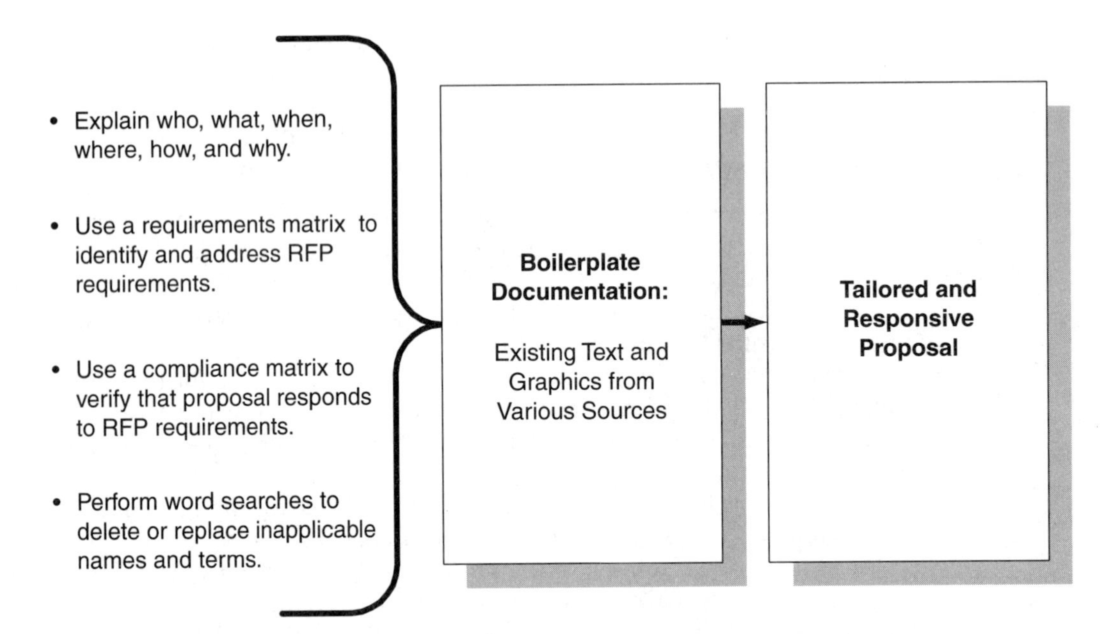

FIGURE 4–3
Tailoring boilerplate

- Carefully check all boilerplate text and graphics to delete or replace inappropriate names or terms used in the source material. If available, use the automated features of your word processing software to help conduct this search.

■ *Proposal Writing Rule #5: Choose the right verb tense to convey the availability of your product or service.*

A customer can be more comfortable about buying a product or service that is available and proven, rather than one that must be developed. The present and future verb tenses can reflect the availability status of your product or service as shown in the following examples:

- Present tense—for something that exists or now occurs ("the engineering services *are available* in 50 states").
- Future tense—for something that will exist or occur in the future ("the engineering services *will be available* in 50 states").

It may be argued that it's always best to use present tense because of its positive message of availability; however, total use of present tense—without appropriate explanation—can confuse readers as they try to differentiate between now and the future.

However, if readers clearly understand that you're referring to something in the future, using the present tense can help the flow of your text and convey the availability image. For example, it could be helpful to use present tense for describing the following: (1) procedures that are established and followed now, but won't be implemented for the customer until the future, (2) proposed milestones for a project schedule that will begin after contract award, or (3) the expected operation of a product that will be delivered to the customer, although it's still in development.

■ *Proposal Writing Rule #6: Substantiate your proposal claims.*

Unsupported claims, superlatives, and platitudes do little to support the credibility of your proposal. Don't expect readers to believe the superiority of your product or service without giving them substantiation. When you make a claim or state a major selling point (proposal theme), back it up with facts.

For example, to substantiate your claim of product reliability, you might describe the reliability in hours of expected usage without major component failure based on documented test results. To substantiate your claim of extensive experience, you might cite the number of years your company has been in business, the experience of your management in average years per person, the number of customers you have served, or the yearly growth of income and revenue since the inception of your company. If you cite experience as substantiation, explain how that experience is specifically related to and will be applied to the proposed work; don't assume readers will make this correlation on their own.

■ *Proposal Writing Rule #7: Sell benefits, not features.*

> You ought to be writing proposals to sell stuff. Products, services, projects, ideas. Whatever you've got. The proposal is a marketing tool; it helps you make money by convincing people to contract with you for the kinds of things you can provide. The proposal positions your product or service as a solution to a business problem.
>
> (Sant, 1992, p. 9)

Customers don't buy features; they buy the benefits provided by the features of your product or service. You buy the saw because it cuts wood, not because it has a brown handle. Don't assume that customers will intuitively understand how a proposal feature will be advantageous to them.

To sell features of your proposed product or service, write to support the themes identified in your storyboard. Amplify the features by clearly explaining how they'll provide real benefits to meet your customers' needs. For example, the benefit might allow the customer to save money, resolve an employee retention problem, receive a much needed product in large quantities within a tight delivery deadline, or receive a safe and reliable product.

■ *Proposal Writing Rule #8: Be logical in your persuasive argument.*

Good argument forms the basis for effective proposal writing. The strongest form of argument—called persuasion—tries to convince the reader to adopt a certain point of view or pursue a certain line of action. Persuasion seeks obvious changes in opinions or action, while the argument presents evidence or logic to support a point of view. As you "argue" why your product or service should be purchased, use logic in substantiating your claims and citing the benefits of what you propose.

Avoid argument fallacies that weaken the credibility of your proposal. Table 4–2 defines seven fallacies and provides an example of each. These examples show "arguments" that might have come from a proposal trying to sell new office chairs to one office location of a multibranch company.

TABLE 4–2: Argument Fallacies

Name	Definition	Example
Ad hominem (Latin: "To the man")	Arguing against a person rather than discussing the issue	Suggesting that the current chairs should be replaced because the purchasing agent who ordered the chairs is surly and incompetent
Circular reasoning	Failing to give a reason something is or isn't true, other than stating that it is	Proposing that the straight-backed chairs should be replaced simply because they are straight-backed chairs
Either/or fallacy	Stating that only two alternatives exist—yours and another one that's much worse—when, in fact, there are other options	Claiming that if the office doesn't purchase new chairs next week, 15 employees will quit
False analogy	Suggesting that one thing should be true because of its similarity to something else, when, in fact, the items aren't enough alike to justify the analogy	Suggesting that the chairs in the office should be replaced because another branch office has many more expensive chairs
Hasty generalization	Forming a generalization without adequate supporting evidence	Proposing that all current office furniture is unacceptable because it was manufactured by a competitor
Non sequitur (Latin: "It does not follow")	Making a statement that doesn't follow logically from what came before it	Claiming that if the office chairs aren't replaced soon, productivity will decrease so much that the future of the office will be in jeopardy—even though there's no evidence that the chairs will have this dramatic effect on business
Post hoc, ergo propter hoc (Latin: "After this, therefore because of this")	Claiming a cause-effect relationship between two events simply because one occurred before the other	Associating the departure of an excellent secretary with the purchase of the poorly designed chairs a week earlier—even though no evidence supports the connection

■ *Proposal Writing Rule #9: Attack your proposal weaknesses head on.*

There's a natural reluctance to avoid placing anything in a proposal that even suggests a weakness in your product or service or your ability to meet customer needs. We believe it's normally best for your proposal to address real or perceived weaknesses candidly. In doing so, acknowledge real weaknesses and explain what you have done to mitigate or eliminate them. For the perceived ones, explain why they never existed. Assume that the readers will know about your weaknesses and take the attitude that it's better to have customers read your side of the issue instead of you hoping they ignore or are unaware of the weaknesses.

The strategy plan, as described in Chapter 1, will help you develop strategies to offset weaknesses. The following are examples of such strategies:

EXAMPLES

- A customer's RFP requirement is too stringent for you to meet.

  STRATEGY: Explain why you believe it is, and then offer and justify a substitute approach that will adequately meet the intent of the requirement. (Regardless, the customer may find your proposal unresponsive because you can't comply with all RFP requirements.)

- Your product has been prone to mechanical failure.

  STRATEGY: Explain how a redesign has eliminated the problem and substantiate the claim with reliability results from independent tests and actual customer usage.

- Your proposed approach has a cost, schedule, or performance risk that seriously jeopardizes your ability to meet customer requirements.

  STRATEGY: Acknowledge the risk and describe what you've done and will do to eliminate it or reduce it to an acceptable level. (This isn't to suggest that all risk must be addressed, because there are different levels of risk—some expected and easily managed. It's the moderate to high risk that deserves your attention in the proposal.)

- The customer has a false perception that your product isn't safe to use.

  STRATEGY: Provide background information about what caused this perception and substantiate that your product is, in fact, safe to use. (This approach can be helpful if you think your product or service has been unjustly criticized by your competition or by a customer decision maker who has a negative opinion about your offering.)

■ *Proposal Writing Rule #10: Write lawfully and ethically.*

Comply with legal and ethical standards as you write the proposal. If the proposal is to include information from copyrighted sources, including the Internet, obtain any required permission before using it and cite its source in your proposal. As previously noted, never put anything knowingly false into your proposal.

You may find that your proposal team has acquired a proposal or parts of a proposal produced by other companies. This information might be provided by your proposal consultants or internal employees, who wrote or obtained it when they were employed elsewhere. Our advice? Refuse to use it or even read it unless you're sure that you can legally do so. Furthermore, if a consultant offers you the unauthorized use of a proposal from another company—regardless if it's one of your competitors or not—refuse the offer and fire the consultant. It's a good bet that if consultants will make that offer to you, they would be willing to make *your* proposal available to other companies. (Also question the integrity of a colleague employee who would offer you the use of unauthorized proprietary information.)

OVERCOMING WRITER'S BLOCK

> When people procrastinate writing, they may be experiencing writer's block. Everyone who writes has experienced writer's block at one time or another, but writer's block is especially critical for proposal writers because proposals almost always have a specific deadline. Those who would write proposals do not often have the luxury of giving in to writer's block. The four major causes of writer's block are insufficient preparation, fear of rejection, fear of producing an imperfect document, and fear of making a wrong decision.
>
> (Bowman and Branchaw, 1992, p. 60)

So far, this chapter has centered on analyzing your readers and writing the proposal to meet their needs. What about the needs of the writer who is faced with a problem that even the best proposal writer can face: writer's block, or the inability to begin or sustain a writing assignment? The following tips can help you start and then keep the words flowing.

■ *Tip #1: Follow your outline and storyboard.*

Use the outline and storyboard to get you moving at the outset—and keep you moving—by listing details that will form the basis of your writing.

■ *Tip #2: Write quickly.*

Force yourself to write quickly. Don't worry about errors in word choice, grammar, or mechanics; just get as much material written as you can. Writers can slow themselves down when they dwell on every word, trying to create perfect copy during the drafting stage. Reserve that penchant for perfection for the editing process (assuming that perfection is ever attainable).

■ *Tip #3: Write in any sequence.*

Don't feel compelled to write in the order the information occurs in the outline. Instead, start writing the body sections that you feel most inclined to write; then later piece the draft together.

■ *Tip #4: If you can't write, at least note what you intend to write.*

If you know what you want to write but don't have all the details or can't adequately express yourself, at least summarize in text what you plan to write and what details you will present. This step can provide you with a springboard for more detailed writing later.

■ *Tip #5: Write overview segments later.*

Write overview segments after you complete the detailed portions of the proposal they summarize. Only after you have fleshed out these portions based on your outline are you ready to write the important beginning and ending sections that are often not outlined, such as the introduction and conclusion. Without the details, it can be very difficult to write an overview. (An exception is the executive summary, which as described in Chapter 6 can be started early.)

■ *Tip #6: Take a break.*

When the words aren't flowing and you feel like you're wasting your time even trying, stop and resume the effort later. A break—for a few minutes or few hours—can do wonders for your creative writing juices. (Unfortunately, because of an impending proposal deadline, your break might be shorter than you would prefer.) You can also polish your writing by going back later to edit and proofread your draft.

■ *Tip #7: Use a thesaurus.*

Use a thesaurus when you want more variety in your choice of words or if you're stymied looking for just the right word to express yourself.

SUMMARY

Your proposal writing must overcome the roadblocks your readers face. These obstacles include impatience, time conflicts, interruptions, lack of knowledge, and shared decision-making responsibilities.

Meet with your customers as much as you can before and during proposal development. Listen to what they say about their needs and how they feel about your proposed offering. Take advantage of their request to provide them with information and comments related to your proposal.

Analyze your readers' needs and write to meet those needs. The following steps will help you plan and conduct an effective reader analysis as you prepare to write:

1. Write down what you first know about your readers.
2. Read proposals written to the same readers or customers; talk to colleagues who developed those proposals.
3. Pinpoint the decision maker(s) in your audience.
4. Determine audience knowledge of the proposal topics.
5. Conduct a strategy meeting.

The effectiveness of a proposal is determined not only by the content of its writing, but also by the way the writing is organized, displayed, and located—known as the principles of organization. The following techniques will help you apply these principles:

1. Write and organize different parts for different readers.
2. Emphasize proposal beginnings and endings.
3. Repeat key points when necessary.
4. Use design techniques to attract reader attention.

The following are basic writing rules:

1. Write with effective style.
2. Use correct grammar.
3. Write with clarity, conciseness, and accuracy.
4. Use paragraphs as the framework of your writing.

The following are rules targeted for proposal writing:

1. Follow your writing plan.
2. Provide the basic proposal details.
3. Ensure your proposal details have purpose.
4. Use "boilerplate" information for your proposal, but use it carefully.
5. Choose the right verb tense to convey the availability of your product or service.
6. Substantiate your proposal claims.
7. Sell benefits, not features.
8. Be logical in your persuasive argument.
9. Attack your proposal weaknesses head on.
10. Write lawfully and ethically.

Overcome writer's block with the following tips:

1. Follow your outline and storyboard.
2. Write quickly.
3. Write in any sequence.
4. If you can't write, at least note what you intend to write.
5. Write overview segments later.
6. Take a break.
7. Use a thesaurus.

EXERCISES

1. Analyzing Reader Needs: In-house Proposal

For this exercise, use any organization to which you now belong—for example, your employment organization, a civic group, or your college or university. Select an idea that you could propose as a member of that organization. Now report on how you would follow the steps in this chapter for determining your readers' needs. This exercise can serve as a basis for a group discussion, an individual oral report, or an individual written report.

2. Analyzing Reader Needs: Sales Proposal

Use nonproprietary data from your own or others' work experience. Assume that you want to write a sales proposal offering a product or service to a new or familiar customer. Now report on how you would follow the steps in this chapter for determining your readers' needs. This exercise can serve as a basis for a group discussion, an individual oral report, or an individual written report.

3. Main Principle of Organization

Use the main principle of organization, as described in Organization Rule #2 (emphasize proposal beginnings and endings), to rewrite the following paragraph:

When our firm was founded 10 years ago, there were just two secretaries to handle all the typing and filing responsibilities for 15 consulting engineers. With their heavy workload, the secretaries needed to resolve questions quickly about the format of documents and office filing procedures. With no official manual available, they simply discussed problems as they encountered them. Frequent communication was the key. Now, however, we have 12 secretaries handling the workload of more than 100 engineers in a four-story building. Without the opportunity to meet frequently, these secretaries often end up using different standards. Clearly, we need a document procedures manual to eliminate this confusion.

4. Principles of Organization and Basic and Proposal Writing Rules

Read the formal proposal example in Chapter 7. Give examples of how this proposal complies with the principles of organization and the basic and proposal writing rules described in this chapter. Make recommendations about how the proposal example could be improved by complying with these rules. This exercise can serve as a basis for a group discussion, an individual oral report, or an individual written report.

5. Principles of Organization and Persuasive and Credible Writing

Select an editorial from your local newspaper that discusses a problem and makes a specific recommendation to solve the problem. How well did it comply with the principles of organization described in this chapter? How well did it describe the problem? Was its recommendation logical, credible, and convincing? How could the editorial be revised to strengthen the validity of its recommendation? This exercise can serve as a basis for a group discussion, an individual oral report, or an individual written report.

6. *Principles of Organization and Basic and Proposal Writing Rules*

Find newspaper or magazine articles that show good and poor compliance with the principles of organization and the basic and proposal writing rules. This exercise can serve as a basis for a group discussion, an individual oral report, or an individual written report.

7. *Proposal Writing Rule: Sell Benefits, Not Features*

Find newspaper or magazine advertisements that show good and poor compliance with Proposal Writing Rule #7 (sell benefits, not features). This exercise can serve as a basis for a group discussion, an individual oral report, or an individual written report.

8. *Proposal Writing Rules: Substantiate Your Proposal Claims and Sell Benefits, Not Features*

Revise your own résumé to comply with Proposal Writing Rule #6 (substantiate your proposal claims) and Rule #7 (sell benefits, not features). Explain how these changes improved your résumé.

This exercise can serve as a basis for a group discussion, an individual oral report, or an individual written report.

Graphics

OBJECTIVES

- **Learn how to determine the graphics needed by your readers**
- **Understand the major reasons to use graphics in proposals**
- **Learn general rules for using all graphics**
- **Learn specific rules for using pie charts, bar graphs, line graphs, schedules, flow diagrams, organization charts, formal tables, drawings, and photos**
- **Perform exercises using chapter guidelines**

"Write to the art," say those who rely mainly on graphics to tell their story, relegating text to a secondary role. "Use graphics only to reinforce the text," say others who maintain the primacy of the words. You need to maintain a balance between these two viewpoints, as you effectively use graphics in your proposals.

This chapter explains the factors that influence the use of graphics in a proposal. It then provides general rules for developing all graphics, followed by rules for developing nine types of common proposal graphics: pie charts, bar graphs, line graphs, schedules, flow diagrams, organization charts, tables, drawings, and photos.

> Pictures sell more cars than words do. More people have photo albums than diaries. . . . The message is clear: Most people would rather look at a picture than read a book. Likewise, most readers would prefer to read an illustrated proposal than a proposal composed entirely of gray pages of text.
>
> (Clauser, 1993, p. 226)

Three points of terminology will help your understanding of graphics in this chapter:

1. The terms *visual aids* and *illustrations* are used synonymously for *graphics* to mean any nontextual part of a proposal. We'll use these three terms interchangeably.
2. Graphics are composed of two subcategories called *tables* and *figures.* Tables organize numbers or words into grids with columns and rows. Figures include all visuals other than tables.
3. Headings, text, and numbers can be used to identify or describe parts of a graphic. We'll use the term *label* for these graphic components.

FACTORS THAT INFLUENCE THE USE OF GRAPHICS

What graphics should you use in a proposal? The answer to this question should be based on the needs of your readers. These needs should be known before you write the proposal. The following are ways to discover these needs:

- Ask in preproposal meetings what kind of graphics the readers most favor or, better, show them samples so they become involved in graphic selection. Your manager may suggest, for example, that an in-house proposal for a new salary matrix contain graphs contrasting present median salaries with proposed median salaries. Or, in a preproposal meeting, a customer may indicate a preference for pie charts and bar graphs to show cost breakdowns for your proposed services.
- Look at the reports and proposals that have been written successfully for the same readers. This use of old reports and proposals is one of the reasons for maintaining a good reference library, available to the proposal team for "borrowing" graphic and text ideas for new proposals.
- Note the readers' use of visual aids in their own reports, brochures, and other documents. What they produce can provide insight into what they wish to receive. For example, if they place illustrations on the cover pages of their reports, they may favor a graphic on the cover page of proposals they receive.
- Carefully read the proposal instructions in the RFP; they may dictate the format and content of specific graphics and limit the type size of graphic labels. Comply with these instructions.

One main goal should drive the use of proposal graphics to make the document more persuasive, informative, and interesting to your readers. However, graphics also help your proposal team become better *writers.* The following are specific reasons for using proposal graphics.

■ *Reason #1: Graphics simplify ideas for the reader.*

Graphics are an ideal way to simplify an idea that requires detailed text for its explanation. They can be used to complement, reduce, or even replace text.

Let's look at an example of how a visual aid can complement detailed text. Assume that the following text appears in your proposal:

> Supertech's innovative Excalibur Abrasive Waterjet can significantly increase your plant's efficiency by replacing conventional cutting tools. The Excalibur makes all cuts with focused, highly pressurized tap water containing abrasive particles. Once combined, the tap water/abrasive mixture accelerates through a high-grade sapphire nozzle and proceeds to hit the work piece at more than twice the speed of sound. Because of this speed, and because the thin super-jet stream is only .018 inch in diameter, the cut produces an exceptional surface finish, and it requires no secondary finishing operation.

By using the cutaway view of the Excalibur, as shown in Figure 5–1, you could quickly help readers understand a process that was described in more than 80 words of text. The graphic would be especially useful for a decision maker who needs to review the proposal quickly.

In addition to simplifying the explanation of ideas, graphics can be used to trim proposal text. For example, instead of using a three-page text narrative to explain a complex manufacturing and assembly process, you might condense this explanation into a series of separate but related tasks summarized in a one-page flow diagram.

■ *Reason #2: Graphics reinforce ideas for the reader.*

> People tend to remember only 10% of what they have read, while they tend to remember 30% of what they have seen. By saying it both in words and in a picture, your audience may remember as much as 50% of your message.
>
> (Woolston, Robinson, and Kutzbach, 1988, reprinted 1990, p. 76)

Use graphics to reinforce main points throughout the proposal text. As an example, assume that you want to persuade a major coastal city to expand its main harbor to accommodate more recreational boats. The following text appears in your proposal:

> A 1988 marketing survey indicated there were 20,000 wet slips available, with a demand for an additional 5,000. However, there is an even more compelling reason to expand. The survey also projected that, in 12 years, there will be 40,000 to 50,000 slips needed. However, with current construction plans, there will be only 25,000 to 35,000 slips available, leading to a significant shortfall.

As you can see by this example, numbers in paragraph form can be difficult to follow and understand. To complement this text, you could capture attention by using a line graph, as shown in Figure 5–2, to reinforce the point about the increasing gap between the supply and demand of wet slips. The figure draws attention to the need for more slips. This emphasis would prime your readers to look carefully at the proposed solution. This is also a good example of a simple graphic helping to support a detailed textual description.

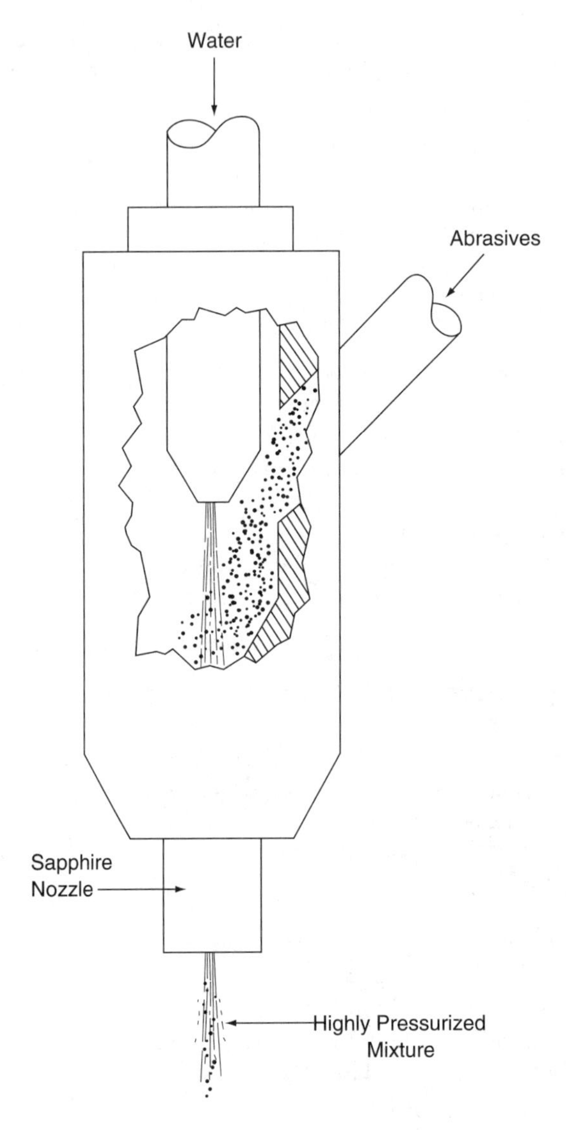

FIGURE 5–1
Cutaway view in a drawing

The type of graphic can convey an idea or a message even if readers don't closely examine its content. For example, a detailed—yet orderly—flow diagram can indicate that you have a sound understanding of a procedure, even if readers don't follow the graphic from beginning to end.

■ *Reason #3: Graphics stimulate reader interest.*

Effective graphics can take precedence over words in the proposal and help "grab" the attention of your readers. The content, design, and artistic quality of graphics

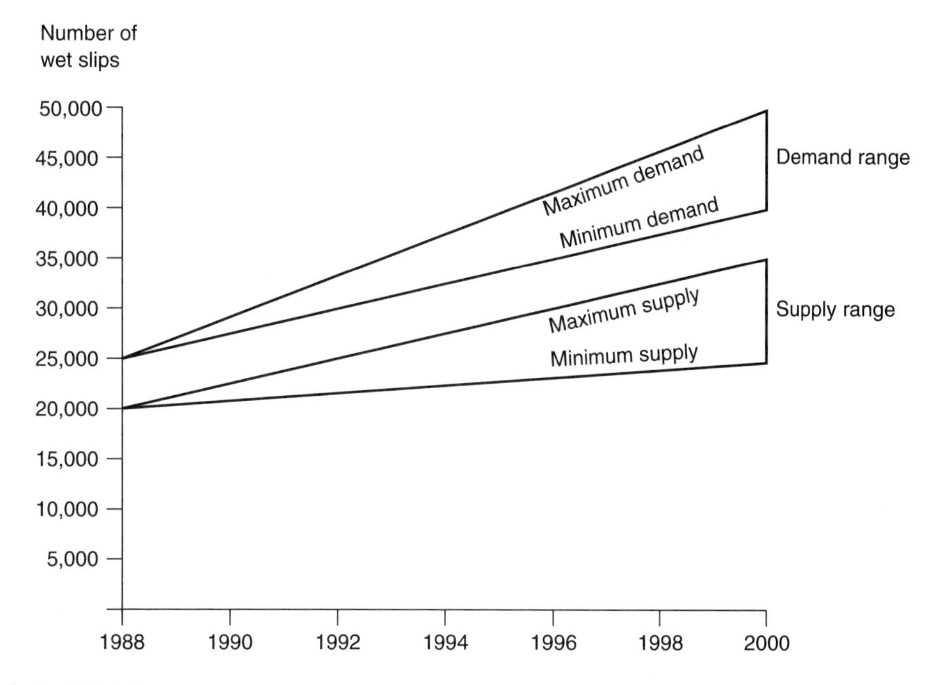

FIGURE 5–2
Emphasizing a need with a line graph

can make your proposal more inviting and interesting to readers, and can lead them to read the text more carefully. The proposal can also be more attractive and easier to read if you use graphics to vary page layout and to break up text.

Graphics that stimulate interest may appear anywhere in the proposal but are particularly effective at the following points:

- Cover of a formal proposal—The cover, including the front, spine, and back, is an effective location for graphics that represent key themes or selling points of your proposal. It's often the first part of your proposal that readers see. Therefore, use this first impression to gain their interest and introduce them to the reasons your proposal should be a winner.
- Title page of a formal proposal—This page is often unimaginative, serving mostly as an administrative aid. Break this pattern and use graphics on the title page to further interest your readers. For continuing emphasis on the proposal themes, you might use all or some of the theme-related graphics that appear on the cover. Also consider the use of color text, reverse type, and text boxes to enliven the appearance of the page. You might even print the title page on colored paper.
- Beginning of a major section and subsections—A graphic at these positions can effectively break your proposal into segments while highlighting key points in the upcoming section. For example, tabs used to separate major sections can include the title of the section, along with a graphic that represents a major point

of the section. (Be careful—if your tab has too much detail, it could be considered a page in a proposal that is page-limited by the RFP. It might be more productive to use that tab "page" for text and graphics in the main body of your proposal.) You could also use a boxed and bulleted list of key points above or next to a section heading.

Used at these points, illustrations can entice the reader to find supporting information in the text. Indeed, a picture can make someone *read* a thousand words.

■ *Reason #4: Graphics help the proposal writer.*

The preceding three reasons for using graphics focus on helping and influencing readers. A fourth reason, although it affects readers, also helps proposal writers. The act of planning and developing graphics causes proposal writers to think through the details of what they want the proposal to say—and this step can improve the content of what they write. The integrated planning of your text and graphics should occur in the storyboard process, as described in Chapter 3.

> Illustrations and drawings should be focused. That is, they should present a single concept. Therefore, they should be clean and uncluttered. Everything not pertaining to the single concept should be eliminated. No detail should be present that does not contribute to the presentation of that single concept.
>
> (Freeman and Bacon, 1990, reprinted 1995, p. 98)

GENERAL RULES FOR GRAPHICS

The following are basic rules that apply to the development of all proposal graphics.

■ *General Graphics Rule #1: Follow a standard approach for development and revision.*

Assign responsibilities for drafting and revising the graphics, and choose the graphics software packages and file formats you'll use. Depending on your labor and material resources, graphics can be drawn by the writers on the proposal team or by illustrators in your graphics or production group. You may prefer to follow a team approach, having authors initially draft the graphics for later refinement by an illustrator. If an art group draws graphics for the proposal writers, review the graphic instructions or drafts before they are passed to an illustrator for work. This review will help you limit (1) the revision of graphics caused by vague instructions and (2) the creation of graphics that are never used because they are poorly designed or unnecessary. Remember that using proposal graphics from other proposals and other reference documents can reduce your graphic workload. If you don't have electronic versions of these reference graphics, consider the use of a scanner to convert the graphics into a digital format. When graphic development begins, follow standard procedures for reviewing and revising graphics and approving them for final layout.

■ *General Graphics Rule #2: Follow style standards for consistency.*

Set and follow style guidelines to promote the uniform display of graphics in your proposal. The guidelines can address the following decisions:

- Allowable image size of the graphics based on the space available in the page layout
- Font choice, capitalization, and type size of the labels
- Width of lines (rules) and borders used with the graphics
- Formats of graphic layouts, especially those for schedules, tables, flow diagrams, and organization charts

■ *General Graphics Rule #3: Use tracking and identification systems.*

You should implement two important administrative systems for your graphics: (1) a system for tracking graphics development and (2) a system for identifying graphics in the proposal.

The tracking system should use a code to track the status of graphics throughout their drafting and revision process. One portion of the code should remain assigned to the graphic throughout its development; another part should change to identify different versions of the graphic so its most current draft can be readily identified. The codes can take many forms. The following example shows one of many possible approaches:

EXAMPLE

Assume that the following basic code numbers are assigned by the proposal coordinator to track graphics in each volume of a three-volume proposal:

- I-001 through I-100 for Volume I graphics
- II-001 through II-100 for Volume II graphics
- III-001 through III-100 for Volume III graphics

(The proposal coordinator was certain that no volume would need more than 100 numbers. However, more numbers could have been issued if required.)

As each graphic is accepted for processing, it receives its own basic code number for the applicable volume. Let's assume a Volume I graphic has been assigned the basic code number of I-001. This graphic retains this basic code during its entire development phase. As it's revised, an extension number is added to the basic code to indicate the revision number of the graphic: I-001-1, I-001-2, etc.

The identification system provides the numbers used with the graphic captions to identify graphics as they appear in the proposal. In contrast to the basic tracking codes, identification numbers may change as the order of graphics changes in a developing proposal. The following are examples of identification number approaches:

EXAMPLES

- Identification numbers can be used to represent the sequential order of graphics as they appear in the entire proposal—for example, Figure 1, 2, 3, etc.
- Identification numbers can be used to represent only first level sections and sequential order of the graphics within that section—for example, Figure 1–1, 1–2, and 1–3 for the first three graphics of Section 1; or Figure II–1, II–2, and II–3 for the first three graphics of Section II.
- Identification numbers can be used to represent any level of sections and sequential order of the graphics within those sections—for example, Figure 2.1–1 for the first graphic in Section 2.1, or Figure 2.1.1–2 for the second graphic in Section 2.1.1.

Use the tracking and identification systems that best meet your needs. When they are in place, you should also use a logging system that allows you to associate the tracking code with the proposal identification number.

General Graphics Rule #4: Refer to all graphics.

All graphics, no matter how self-explanatory, should be supported by references in the proposal text. The following are guidelines for writing text references to your graphics:

- Refer to the graphic by its identification number.
- Incorporate the reference smoothly into the related text discussion, summarizing the major point of the graphic.
- Refer to the caption and even the page number of the graphic, if they're needed for clarity and emphasis. (However, these identification features aren't usually needed. If they're used, remember that they can take up valuable space in a page-limited proposal. Plus, if you refer to the page location, a late revision in your proposal can change its pagination and force you to verify that the page references for graphics are still correct.)

The following are examples of text references to the same graphic:

EXAMPLES

Example 1: In the past three years, 100 midsize electronics firms have shifted to the Accu-count System for recording warehouse inventories. The change to this system has cut losses dramatically, as shown in Figure 3–1 ("Accu-count Decreases Warehouse Losses").

Example 2: As shown in Figure 3–1, the 100 firms that have shifted to the Accu-count System in the last three years have cut their warehouse losses dramatically.

CRITIQUE: Both examples refer to the graphic identification number, and both incorporate the reference smoothly in the discussion, but the examples differ in two ways:

- Example 1 includes the graphic caption; Example 2 doesn't provide it.
- Example 2 mentions the graphic number at the beginning of the passage; Example 1 mentions it at the end.

This second difference deserves more attention. An early reference to the graphic gives it high visibility, but this placement can send readers searching for the graphic before they read about its significance—a response you may or may not want. (It may be desired, for example, if the text describes a series of procedures. It could be very helpful for the reader to see a flow diagram of the overall process first before reading the procedural details.) In contrast, late reference to the graphic allows readers to grasp the importance of the illustration before they turn to it. Choose the option that best suits your purpose.

General Graphics Rule #5: Place graphics near the first reference.

To help readers find and use graphics, display a graphic on the same page of its first textual reference or the nearest page that follows this first reference. However, the following are exceptions to these general rules:

- Visuals referred to throughout a section or an entire proposal can be placed at a central point for ease of reference, such as at the end of a section or in an appendix or attachment at the end of the proposal.
- Less important visual aids can be placed in an appendix or attachment pages at the end of the proposal, where they won't interrupt the flow of text. (The RFP may note that pages in these sections will not be tallied as part of the proposal length in a page-limited proposal. Therefore, placing graphics in these parts of the proposal can help you comply with customer-imposed page limits.)

If text would be better understood by frequent reader glances to a related graphic, such as a table or a flow diagram, place the related text and graphic to allow the reader to view them both without turning a page. If the pages are laid out front to back (both sides of the page), place the text on the left facing page with the graphic on the right facing page. If the pages are laid out on one side only, try to have the related text and graphic on the same page.

General Graphics Rule #6: Strive to arrange graphics to be read without turning the proposal.

Visual aids become a more integral part of the proposal when they can be read without repositioning the document. Therefore, avoid the use of full-page landscape graphics that force readers to turn the proposal sideways to view the graphic.

If feasible, revise a landscape graphic so it can be displayed in a portrait layout. If a landscape graphic must be used, display it on the page lengthwise so that readers will turn the proposal clockwise to see it.

■ *General Graphics Rule #7: Design graphics for readability and comprehension.*

Avoid labels that are written vertically, diagonally, or in all capital letters, because they can be hard to read. To also aid readability, lay out your graphics to accommodate the reading pattern of your typical reader. The normal pattern for U.S. English-speaking readers is to read text and graphics from left to right and from top to bottom.

It's especially important to follow this pattern when making flow diagrams that illustrate the steps of a functional process. If you do, it will help your readers find the beginning point of the process and then follow the orderly progression of the functional steps. Figure 5–3 shows examples of how graphic flow diagrams should be constructed to support normal reader scan patterns.

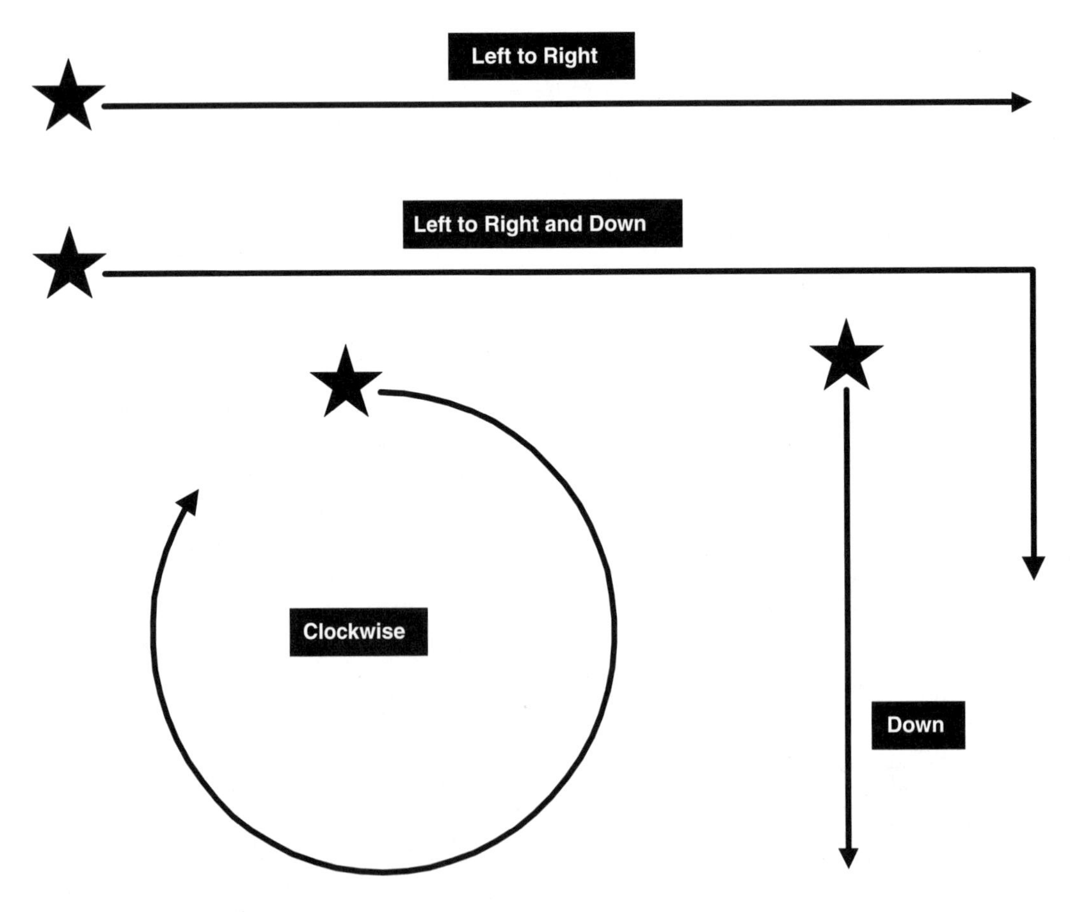

FIGURE 5–3
Scan patterns for graphics layout

■ *General Graphics Rule #8: Avoid clutter.*

Graphics should simplify ideas, reinforce major points, and attract interest. To meet these goals, don't put unnecessary detail in your graphics—specifically, omit labels and other details that aren't immediately relevant to your purpose and the readers' understanding of the graphics. However, don't make your graphics so simple that they convey an ambiguous meaning or have no meaning at all.

Despite your best efforts, you may be forced to violate this rule when an RFP requires great detail in the proposal, but doesn't authorize a reasonable page allowance to package it. To most effectively use the limited space that's available, you may then have to include many details in your graphics to provide the required content.

■ *General Graphics Rule #9: Provide support information.*

Every graphic, whether a formal table or a figure, should have an identification number and caption. These items can be centered or left justified above the tables and either above or below the figures. To make them stand out, you can boldface and italicize these numbers and captions or display them in a type size or font style different from that of the text.

To get the most impact with your captions, use what is known as "action" captions. These captions follow the approach of a newspaper or magazine headline, which summarizes a key point in the related article. Instead of using a caption only as a title, an action caption states a major thrust of the graphic or the conclusion you want the reader to reach after viewing the graphic.

The following examples compare a title caption with action captions for the same graphic:

EXAMPLES

An organization chart of your company's management team for a proposed project could be captioned "Figure 3. XYZ Management Organization Structure"—a useful *title* but one that says nothing about the management structure or approach.

Instead, an *action* caption could identify the graphic with one of the following statements:

- "Figure 3. The XYZ Project Team Is Led by a Single-Point Project Manager with Direct and Responsive Access to Our President and Corporate Resources."
- "Figure 3. Our Management Structure Integrates Our Customers and Subcontractors into the Project Team for More Effective Cost, Schedule, and Performance Control."

Notes, keys, and source data are sometimes needed for your graphic to provide further explanations or to acknowledge borrowed information. Notes contain general pieces of information that put the visual aid into a context and are placed conspicuously beneath the caption. Keys give the meaning of abbreviations or terms in the graphic.

Source data appear at the bottom of the graphics and cite the person, organization, or publication from which the graphic was borrowed. This item deserves special emphasis. You have an ethical and in some cases, a legal obligation to cite the sources from which you borrow information for an illustration. This guideline holds true whether you take an entire visual from a source or adapt it to suit your needs. Your obligation often goes beyond simply citing the source. If you borrow graphics—particularly for an external sales proposal—from a copyrighted publication, request written permission from the source publication to use the information. Because you stand to make a profit as a result of the proposal text, written permission is a must.

It's better to err on the side of excessive notation of sources. There are many books available about citing sources in research and technical writing—use them or consult with a local librarian if you have any doubts about the appropriateness of your citations.

■ *General Graphics Rule #10: Use color for a purpose.*

Avoid the indiscriminate use of color in your graphics. Justify use and expense of color based on a clear need, not just to make a graphic more attractive. For example, it might suffice to use a black-and-white photo of your manufacturing facility instead of a color one. However, you might need color to contrast the flow of conditioned air, electrical power, fuel, oil, and hydraulic fluid in a detailed systems schematic for a military jet fighter.

The overuse of color can be distracting to the reader, who may perceive color was used just because it was available. Advances in publication software and hardware have made it easier and less expensive to create colored graphics. However, the processing, printing, and binding of proposal pages with color can still add extra labor and material costs during document production. Some customers, particularly U.S. government agencies, may discourage the elaborate use of color in the proposal and see its presence as a lack of concern for cost control.

■ *General Graphics Rule #11: Use typographical and design features as graphic components.*

> You can go quite a distance toward making your pages appealing and readable with nothing more than text, headlines and the appropriate manipulation of white space. But the desire to do more, to dress up those pages, can often be overwhelming. To satisfy this urge, the designer adds diagrams, drawings and photographs, while the typographer looks to the rich repertory of pictorial shapes found in typeface designs to add finishing touches.
>
> (Burke, 1990, p. 87)

The creative use of typography and design elements can enhance the graphical image of your proposal. Figure 5–4 shows typography and design features that can be used with your graphics, text, and page layout.

Drop cap

This is an example of a drop cap. It can add a creative touch to the beginning of major sections in your proposal.

Emphasis

These are examples of different types of EMPHASIS in your text. Emphasis can help the reader notice *important* points. Set criteria for its use and **don't** overdo it.

Tables

This an example of a 3-D table.

Column A	Column B
xxxxxx	xxxxxx
xxxxxx	xxxxxx
xxxxxx	xxxxxx
xxxxxx	xxxxxx

This is an example of a table with reverse type on a black background and alternating shaded columns.

Reverse type			
xxxxxx	xxxxxx	xxxxxx	xxxxxx
xxxxxx	xxxxxx	xxxxxx	xxxxxx
xxxxxx	xxxxxx	xxxxxx	xxxxxx

Shaded Border

This is an example of text that is boxed with a shaded border.

Horizontal Rules

This is an example of horizontal rules being used to separate text from other parts of the document. As shown, the rules can be made with different point size.

Vertical Rules

This is an example of vertical rules. Vertical rules can also identify changes to revise a proposal that has already been submitted to the customer.

Bullets, Symbols, and Numbers for Lists

◆ xxxx	◇ xxxx	* xxxx	⇒ xxxx	✔ xxxx	❑ xxxx	① xxxx
◆ xxxx	◇ xxxx	* xxxx	⇒ xxxx	✔ xxxx	❑ xxxx	② xxxx
◆ xxxx	◇ xxxx	* xxxx	⇒ xxxx	✔ xxxx	❑ xxxx	③ xxxx
◆ xxxx	◇ xxxx	* xxxx	⇒ xxxx	✔ xxxx	❑ xxxx	④ xxxx

FIGURE 5–4
Typography and design features

- Rules—These are lines of varying widths and orientation that can be used as design elements in graphics or a page layout. Rules can be single or multiple lines with different widths and orientations (such as vertical, horizontal, or diagonal). They can divide a graphic into segments or separate a series of related graphics. Used in a page layout, rules can divide text columns and separate graphics, headers, footers, and headings from the text.
- Boxes—These design features enclose text or graphics within borders. A box can enclose text or graphics with lined borders, lined borders with outside shading (also known as a drop shadow), or a shaded or tinted area without lines. Within a shaded or tinted area, you can vary the darkness (weight) of the type or reverse the type image by displaying white text on a dark background. Boxes can be used to highlight a key point in the text or draw attention to a key part of a graphic.
- Drop or initial caps—This feature uses oversized capital letters to begin the sentence of a major section or paragraph. In addition, attractive design elements can be added to drop caps to give them a more distinctive appearance.
- Miscellaneous type features—The use of underlines, boldface, capitalization, and italics can add emphasis to text, headings, labels, and captions. Don't dilute their effectiveness by overuse.
- Three-dimensional and perspective imaging—Three-dimensional imaging can be applied to graphics to show depth and add artistic variety. Word processing and graphics software can make it easy to use three-dimensional features, particularly in tables, bar graphs, pie charts, and flow charts.

SPECIFIC RULES FOR NINE COMMON GRAPHICS

In the rest of this chapter, we'll describe rules for using nine types of graphics—pie charts, bar graphs, line graphs, schedules, flow diagrams, organization charts, formal tables, drawings, and photos.

Rules for Pie Charts

Pie charts can gain a reader's attention because they offer a comforting simplicity to even the busiest, most frazzled reader. However, pie charts are so deceptively easy to use that you must be aware of what they can and can't do. They show relationships between the parts and the whole quite well, but they provide only approximate, not exact, information. Moreover, they can confuse readers when they include too many divisions (wedges). To exploit pie charts in your proposal, use the following rules.

■ *Pie Chart Rule #1: Use no more than six or seven wedges.*

The pie chart's selling point is its simplicity. If you clutter it up, the reader can have trouble grasping relationships. Therefore, limit the number of pie chart wedges to no more than six or seven. In fact, the fewer wedges the better. Figure 5–5 conveys

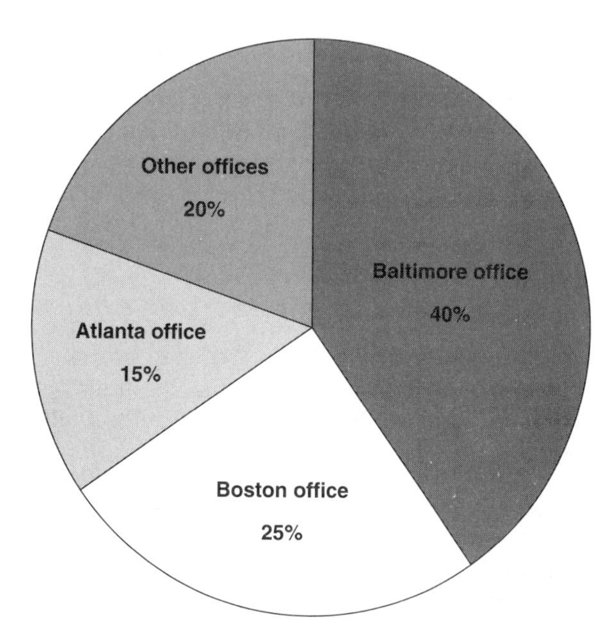

FIGURE 5–5
Pie chart with percentages

a simple, yet informative, message as it shows how proposed work will be divided among various company offices.

■ *Pie Chart Rule #2: Start clockwise at 12:00, moving from the largest to the smallest wedge.*

Make your pies more predictable by matching the expectations of the readers' eyes. The eye follows a pie chart much as it reads a clock—clockwise from the top position. As shown in Figure 5–5, starting at the straight-up position gives more visibility to the Baltimore office, which will provide the most employees (40 percent) for the proposed project. Next in clockwise order comes the Boston office (25 percent), followed by the wedge representing the Atlanta staff (15 percent).

Although this rule dictates a structure that presents wedges in a clockwise order from the largest to the smallest wedges, break the rule if a change in order suits your needs. For example, the Figure 5–5 wedges are shown in a clockwise order of size until the last piece, "Other Offices," which is larger than the preceding wedge. The objective in this pie chart is to draw the most attention to the workforce contribution, in descending order, of the three main offices, Baltimore, Boston, and Atlanta. A grouping of the other six offices is left for the last wedge, even though its percentage total is a bit more than the wedge that precedes it.

■ *Pie Chart Rule #3: Use a wedge scale divisible by 100.*

Pie charts best catch the readers' eyes when they represent items divisible by 100—as with percentages and dollars. Figure 5–5 is an example using percentages. Figure 5–6 shows the average paycheck deductions for company employees as a function of the parts of each dollar deducted. Ensure that your percentages or cents add up to 100.

■ *Pie Chart Rule #4: Be creative but simple.*

Pie charts should be simple in form and content; the basic circle can accommodate only limited detail or visual effects. If you make them too fancy, just as if you add too many wedges, they can lose appeal to readers who are searching for a visual they can comprehend quickly.

Despite these limitations, the following creative features can be used for your pie charts:

- Shading—Darkening a wedge gives it prominence, particularly if it's not the largest one.

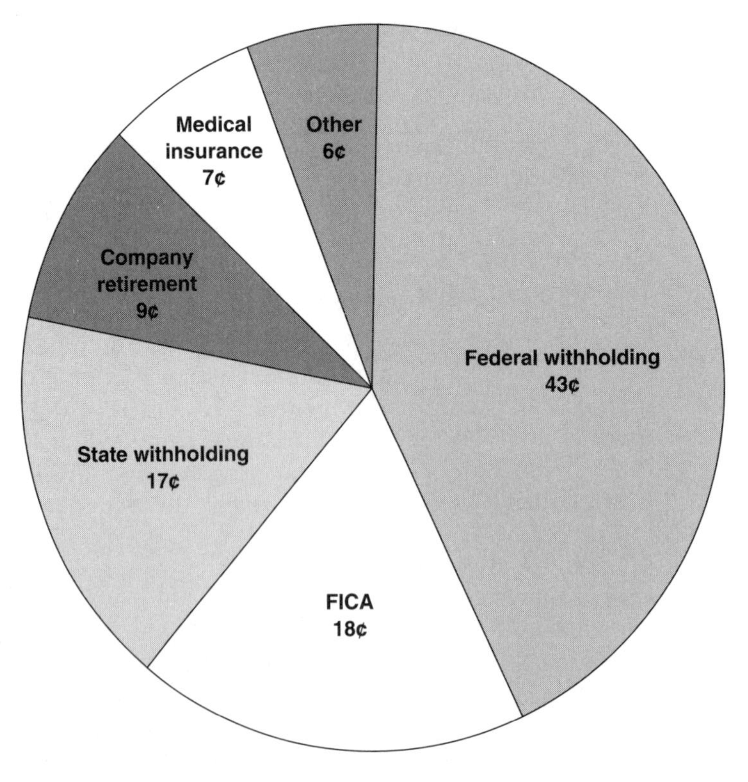

FIGURE 5–6
Pie chart with money

- Cutouts—Separating a wedge from the pie chart, as if a piece of pie were being removed, draws attention to this wedge.
- 3-D—Adding a third dimension to your pie chart gives more life to the drawing.
- Pie within a pie—Amplifying the content of a specific wedge can provide the reader with more information.

Examples of these features are shown in Figure 5–7.

■ *Pie Chart Rule #5: Draw and label carefully.*

The most common pie chart errors are (1) wedge sizes that don't correspond correctly to percentages or money amounts and (2) wedge sizes that are too small to accommodate the information placed in them. The following are guidelines to avoid these mistakes:

- Size—Make sure the chart occupies enough of the page. On a standard 8 1/2″ × 11″ sheet with only one pie chart, your circle should be from 3″ to 6″ in diameter—large enough not to be dwarfed by labels but small enough to leave adequate white space in the margins.
- Labels—Label the wedges either inside or outside the pie, depending on the length of the labels, the number of labels, and the number of wedges. Choose the option that makes the chart easiest to understand.
- Percentage conversion—If you're drawing a pie chart by hand (not using a graphics software program) draw accurate wedges with a protractor or similar device. One percent of the pie equals 3.6 degrees (360 degrees / 100 = 3.6 degrees). With that as your guide, you can convert percentages or cents to degrees. In contrast, the accuracy and speed of software programs can make this conversion an easy task.

Rules for Bar Graphs

Bar graphs provide comparisons by means of two or more bars running either vertically or horizontally in the graphic. They can contain more technical detail than pie charts. The following rules will help you produce effective bar graphs.

■ *Bar Graph Rule #1: Use a limited number of bars.*

Though bar graphs can show more information than pie charts, both types of graphics have their limits. Too many bars can make it difficult for readers to understand the variables depicted in the graphic. However, more bars can be used effectively when the graph is larger.

■ *Bar Graph Rule #2: Show comparisons quickly and accurately.*

Bars should convey information quickly and accurately. Avoid using bars that are too similar in length because it can be difficult to quickly compare measurements represented by these bars.

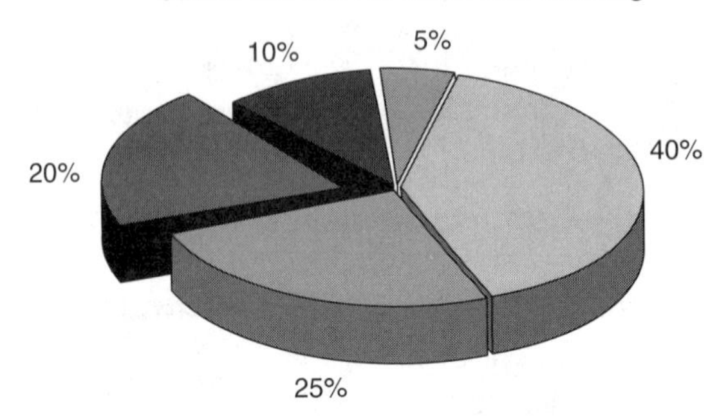

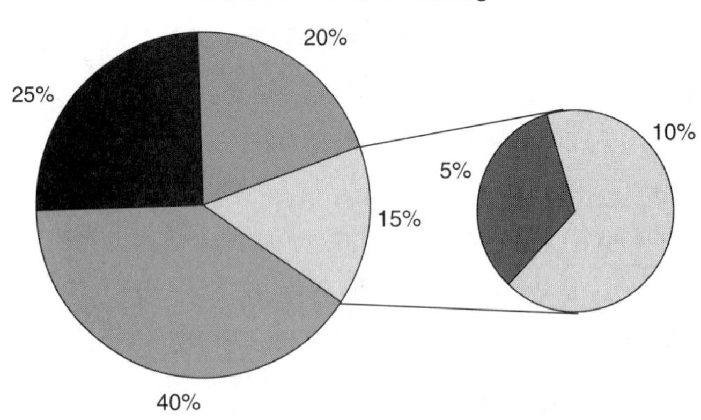

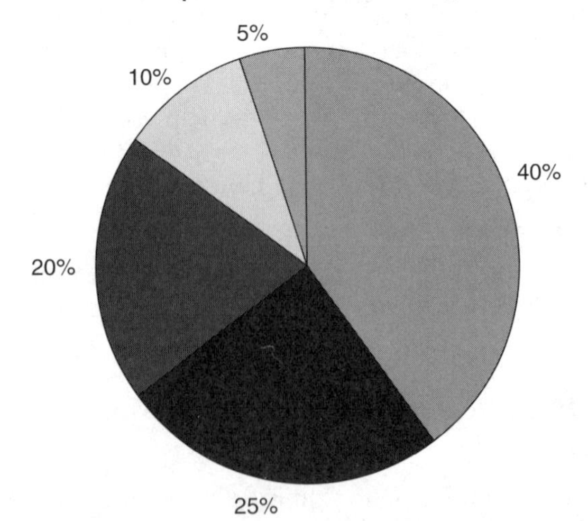

FIGURE 5–7
Creative pie charts

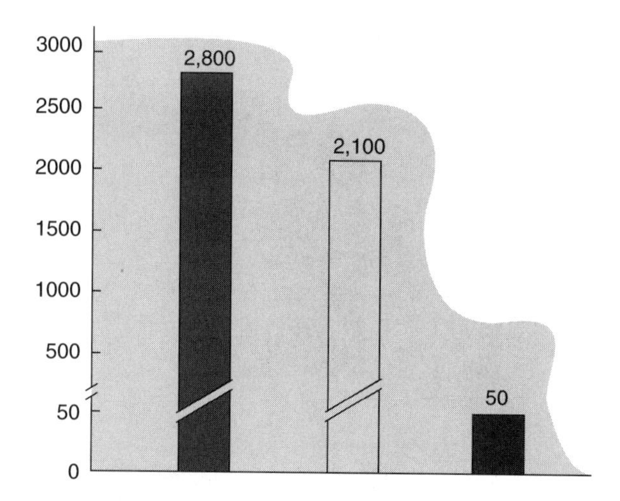

FIGURE 5–8
Misleading break lines in a bar chart

Also avoid the use of "break lines" (two parallel lines) to shorten bars and their associated scales. The technique, as demonstrated in Figure 5–8, is visually deceptive even though the lines indicate where the bar and scale breaks occur. For example, note that Figure 5–8 provides no accurate visual relationship between the "50" bar and the other two bars, because the "2,800" and "2,100" bars have been artificially shortened. If these breaks aren't noticed by readers, they could incorrectly conclude that the relative difference between the "50" bar and the other bars is much less than it really is. To portray the true visual comparison among the three bars, the "2,800" and "2,100" bars must be significantly longer.

In addition, don't begin the horizontal or vertical scales of a bar graph with a number greater than zero (known as the "suppressed zero"). With this scale, bar differences—big or small—can also be exaggerated and misleading.

■ *Bar Graph Rule #3: Vary bar spacing; maintain bar width.*

While bar length and space between bars can vary, keep the bar width constant. The options, illustrated in Figure 5–9, include the following:

- Option A—Use no space when there are close comparisons or many bars, so that differences are easier to grasp.
- Option B—Use equal space, but less than the bar width, when bar length differences are great enough to be seen despite the distance between bars.
- Option C—Use variable space when gaps between some bars are needed to reflect gaps in the data.

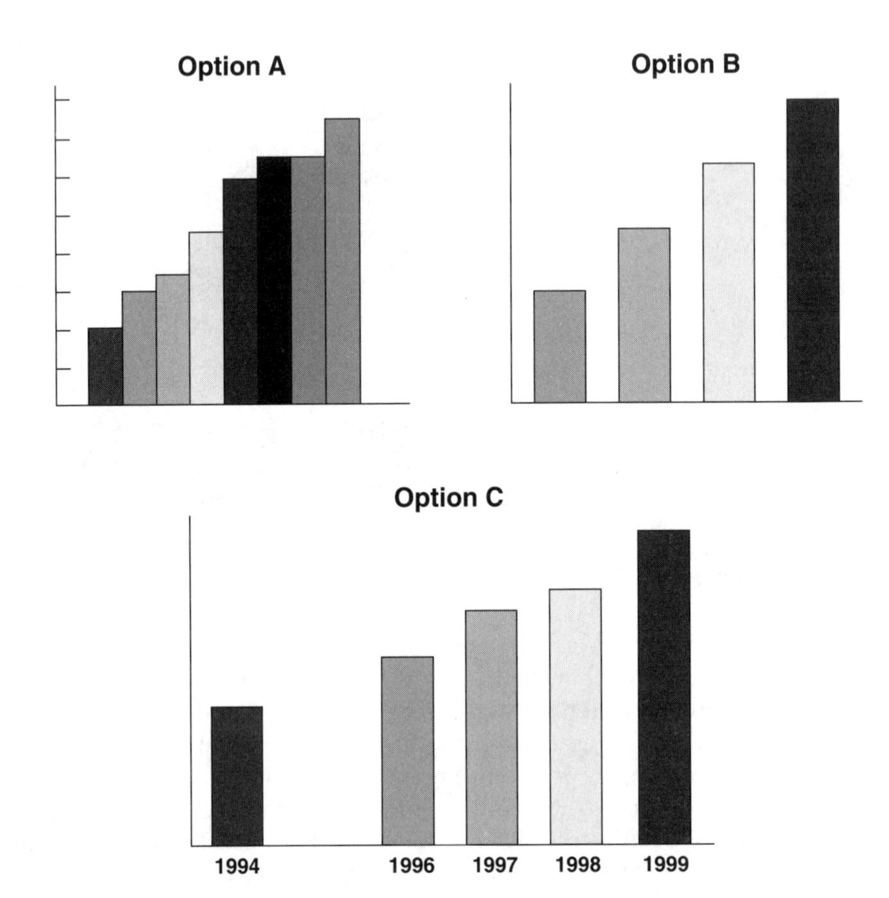

FIGURE 5–9
Spacing options for bar charts

Although white space helps draw attention to the bars, space gaps wider than the bars can take visual emphasis away from the bars themselves.

■ *Bar Graph Rule #4: Arrange bars in an order that best suits your purpose.*

Because readers grasp bar chart meaning by comparing bar lengths, choose an order that meets your objective. The following are approaches for arranging the left to right order of the bars:

- Sequential order—Use when the order of the bars reveals a trend—for example, the increasing number of company projects over the past five years.
- Ascending or descending order—Use when you want readers to be influenced by the rising or falling of the bars—for example, the use of bars showing the fuel efficiency (miles per gallon—mpg) of five compact cars from the lowest to highest, with your firm's car shown by the last bar as the most fuel efficient.

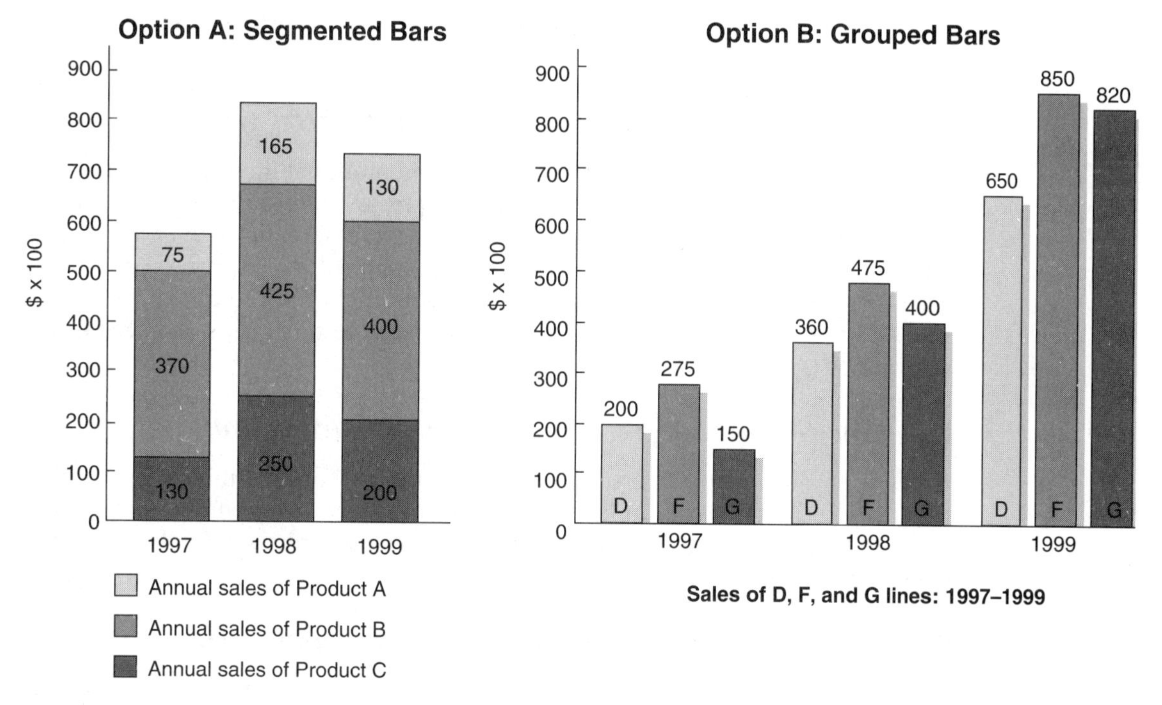

FIGURE 5–10
Showing comparisons with bar charts

- Alphabetical order—Use as a last resort when trends are unimportant and when no specific order serves your purpose—for example, unemployment rates shown for seven Georgia counties displayed in alphabetical order.

Bar Graph Rule #5: Creatively show comparisons.

Use features to creatively show comparisons. For example, the segmented (divided) bars, shading, and grouped bars shown in Figure 5–10 permit the bar graphs to display effective comparisons. Remember not to complicate a bar graph so that you obscure the main comparisons.

Rules for Line Graphs

A line graph can convey meaning quickly. A line that moves up or down, either gradually or sharply, can evoke reader response even before the labels are read. With that potential for impact, the line graph should be an important part of the proposal writer's persuasive strategy.

Line graphs contain vertical and horizontal axes that reflect quantities of two different variables. The vertical (y-axis) usually plots the dependent variable, while the horizontal (x-axis) plots the independent variable. (The dependent variable is affected by changes in the independent.) Lines then connect the points that have been plotted on the graph. The following are rules for planning and drawing line graphs.

■ *Line Graph Rule #1: Use line graphs to stress trends.*

The direction and angle of graph lines carry an emotional message to the reader. Exploit this feature in your proposal. For example, assume you're proposing a new medical plan to a firm that has had double-digit increases in its plan costs over the past five years. Your plan guarantees no increases for the next three years. To influence readers concerned with rising costs, you might use the line graph in Figure 5–11 to show that your plan doesn't project any cost increases over that three-year period.

■ *Line Graph Rule #2: Place line graphs strategically.*

Given their powerful impression, line graphs deserve high visibility in your proposal. They can be placed in key parts of your proposal to serve as an attention-grabber, to emphasize the immediacy of a problem, to stress a major benefit, or to substantiate a claim.

■ *Line Graph Rule #3: Strive for accuracy and clarity.*

Ensure that your line graphs display data accurately and clearly. As with the bar graph, use a scale that doesn't mislead the reader with visual tricks. Exaggerated

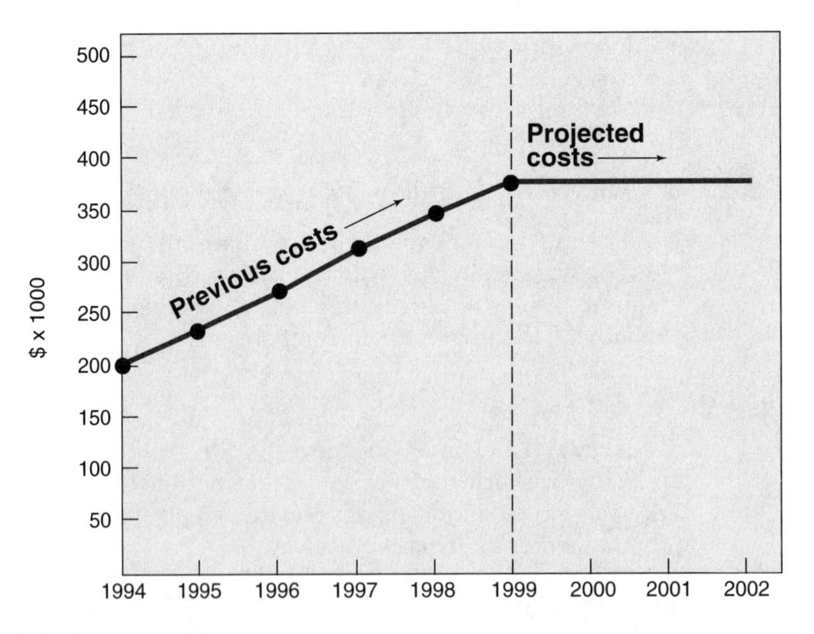

FIGURE 5–11
Projecting a trend with a line graph

changes on line graphs aren't merely inaccurate; they're unethical. The following are ways to maintain both accuracy and visual clarity in line graphs:

- Start all scales from zero to eliminate the possible confusion of a "suppressed zero."
- Select a vertical-to-horizontal ratio for axis lengths that's pleasing to the eye (an effective choice is a vertical axis that measures three-fourths the length of the horizontal axis).
- Make graph lines as thick as, or thicker than, the axis lines.
- Use shading under lines, if this makes the graph more readable.

■ *Line Graph Rule #4: Avoid including numbers on the line graph itself.*

Line graphs rely on simplicity for their effect. Avoid cluttering the graph with numbers that detract from its visual impact.

■ *Line Graph Rule #5: Use multiple lines with care.*

A line graph can, of course, display more than one line. Several lines, as shown in Figure 5–12, allow you to compare and contrast different trends based on the same two variables. Adding too many lines on a graph can confuse the reader with too much data. To avoid this problem, use no more than four or five lines in a single graph. To differentiate among the lines, use different line styles, shades, or colors.

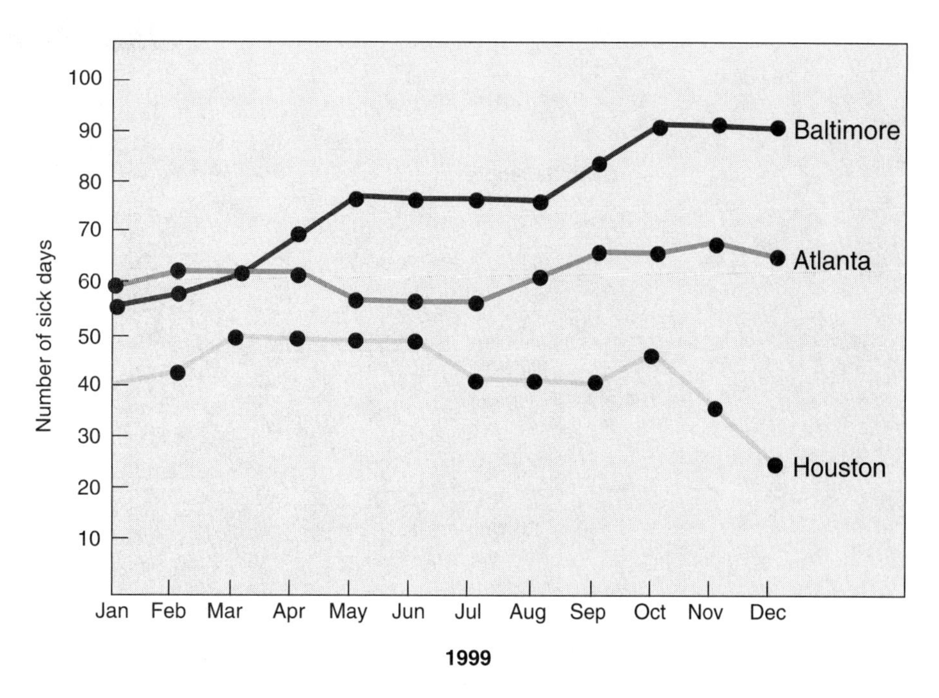

FIGURE 5–12
Using multiple lines in a line graph

Rules for Schedules

There's an opportunity in most proposals to include a schedule that shows when a proposed action will occur. Although the proposal text should describe the activity in detail, an easy-to-read schedule can strongly support the text. Two common types of schedules are the Gantt schedule and milestone schedule, examples of which are shown in Figure 5–13. These schedules include three parts:

- A vertical axis to list the various parts of the project in a sequential order
- A horizontal axis to register the most appropriate time scale
- Horizontal bar lines (Gantt schedule) and/or separate markers (milestone schedule) to show the time needed to complete each task

Use the following rules to produce schedules that are easy for even the most hurried reader to understand.

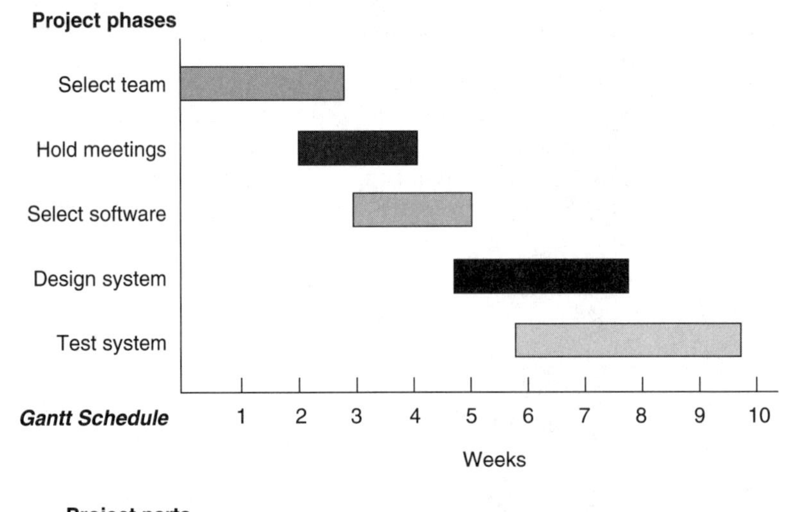

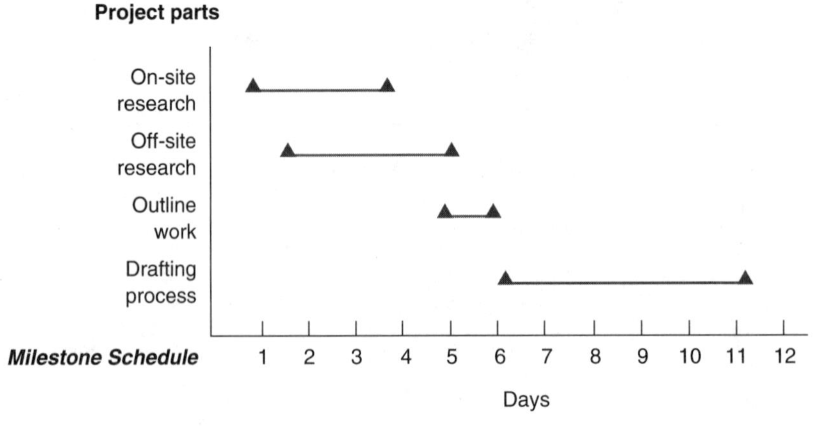

FIGURE 5–13
Gantt and milestone schedules

■ *Schedule Rule #1: Show only main activities and use an appropriate time scale.*

Avoid excessive clutter in the schedule by identifying no more than 10 to 15 main activities. If more detail is needed, construct a series of schedules linked to a main summary schedule.

Use the most appropriate time scale to show when tasks will occur. For example, you can use a calendar or notional time line based on a daily, weekly, monthly, quarterly, or yearly scale. A calendar scale identifies events with a specific date; a notional scale identifies events scheduled in reference to a particular event—whenever it occurs. The notional scale is useful (1) when it's more important to show the progression of activities than to show when they'll occur on the calendar or (2) when you don't have specific dates for the activities.

■ *Schedule Rule #2: List activities in sequence, starting at the top.*

The convention is to list activities in sequence from the top to the bottom of the vertical axis. This arrangement allows the readers' eyes to move from the top left to the bottom right of the schedule.

■ *Schedule Rule #3: Run all labels in the same direction.*

Schedule labels should be displayed in the same direction to have immediate and clear impact. If you force readers to turn the page sidewise and then back, they may lose interest in the schedule.

■ *Schedule Rule #4: Creatively develop different formats.*

Using the Gantt and milestone schedules as your starting point, you can devise many variations to suit your needs. In Figure 5–14, features of the Gantt and milestone schedules are combined to create another format.

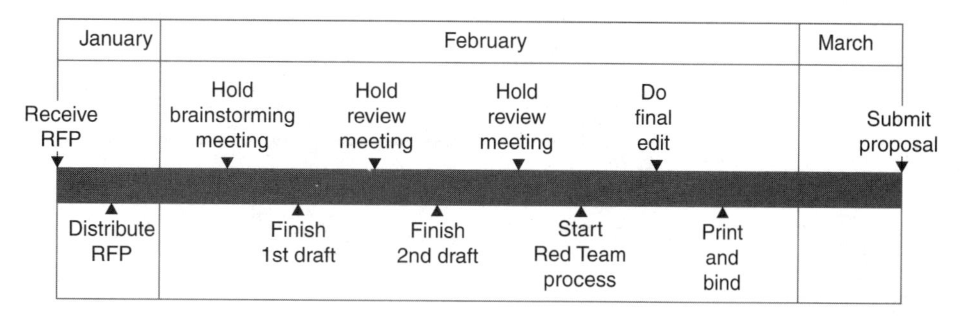

FIGURE 5–14
Hybrid Gantt and milestone schedule

■ *Schedule Rule #5: Be realistic in your schedule.*

Develop schedules that realistically reflect milestone dates and the time required to complete those milestones. Because the proposal may be considered a contract, you could be held accountable to the schedule you offer. If the schedule is based on preliminary planning and will be updated later, indicate that in the proposal.

Rules for Flow Diagrams

A flow diagram tells a story about a process, usually linking a series of boxes or other shapes representing various actions. It can be used to describe any process with defined and interrelated steps. Figure 5–15 shows an example. A flow diagram should simplify the text narrative, not duplicate its complexity. Therefore, keep your flow diagrams short and simple—but not so short and simple that they convey little or no information. The following rules will help you meet those goals.

■ *Flow Diagram Rule #1: Present only overviews.*

Readers look to the flow diagram for the "big picture." Reserve intricate details for the text, appendices, or attachments.

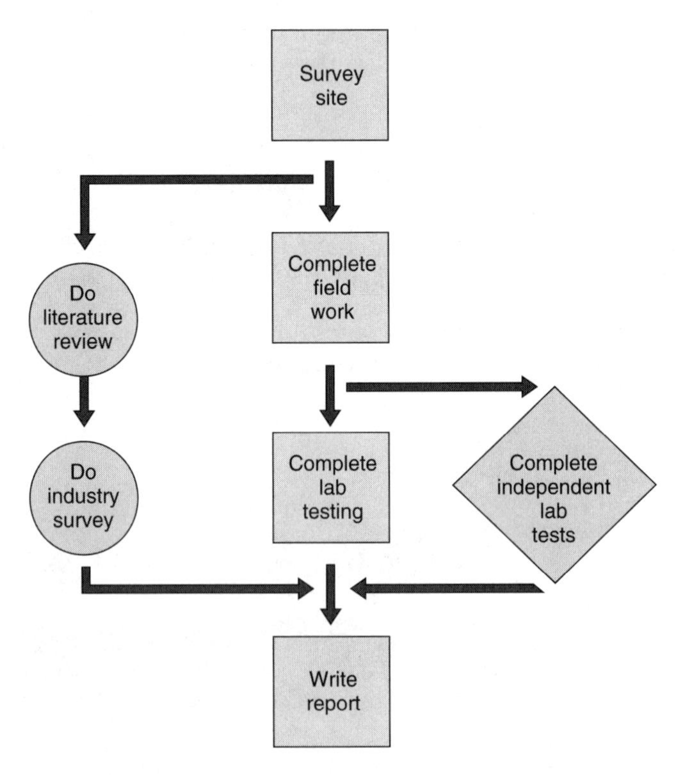

FIGURE 5–15
Showing a process with a flow chart

■ *Flow Diagram Rule #2: Limit the number of shapes.*

Flow diagrams rely on rectangles and other shapes to represent parts of the process. Different shapes should represent different types of activities. This variety helps in describing a complex process, but it can also produce confusion. For the sake of clarity and simplicity, limit the number of different shapes in your flow diagrams.

■ *Flow Diagram Rule #3: Provide a legend that defines the shapes.*

When the flow diagram uses a variety of shapes, include a legend that identifies the meaning of each shape. Don't assume that readers will intuitively understand their meaning.

■ *Flow Diagram Rule #4: Use a sequence to accommodate reader scan patterns.*

Arrange your flow diagrams to comply with General Graphics Rule #7. This approach will help readers follow the sequential order of the displayed process.

■ *Flow Diagram Rule #5: Label all shapes.*

Each shape within a flow diagram should have an associated label. Follow one of the following approaches:

- Place the label inside the shape.
- Place the label immediately outside the shape.
- Insert a reference number in each shape and place a legend for all numbers in another location (preferably on the same page).

The label can identify an object, an action, or a decision represented by the shape, or it can show an input or output between the shapes. Be consistent in your labeling by using parallel wording—for example, to clearly define action in a series of shapes, it is parallel form to label each shape with a verb phrase.

Rules for Organization Charts

Organization charts show the structure of a company or other organization. They use shapes and connecting lines to show relationships among people, positions, responsibilities, and work units. Use the following rules for producing organization charts.

■ *Organization Chart Rule #1: Use shape placement and features to show organization levels.*

The placement and features of the chart shapes can reflect the authority and responsibility of people, positions, or work units. The placement of shapes higher in the chart indicates a higher level of authority or responsibility. Figure 5–16 is

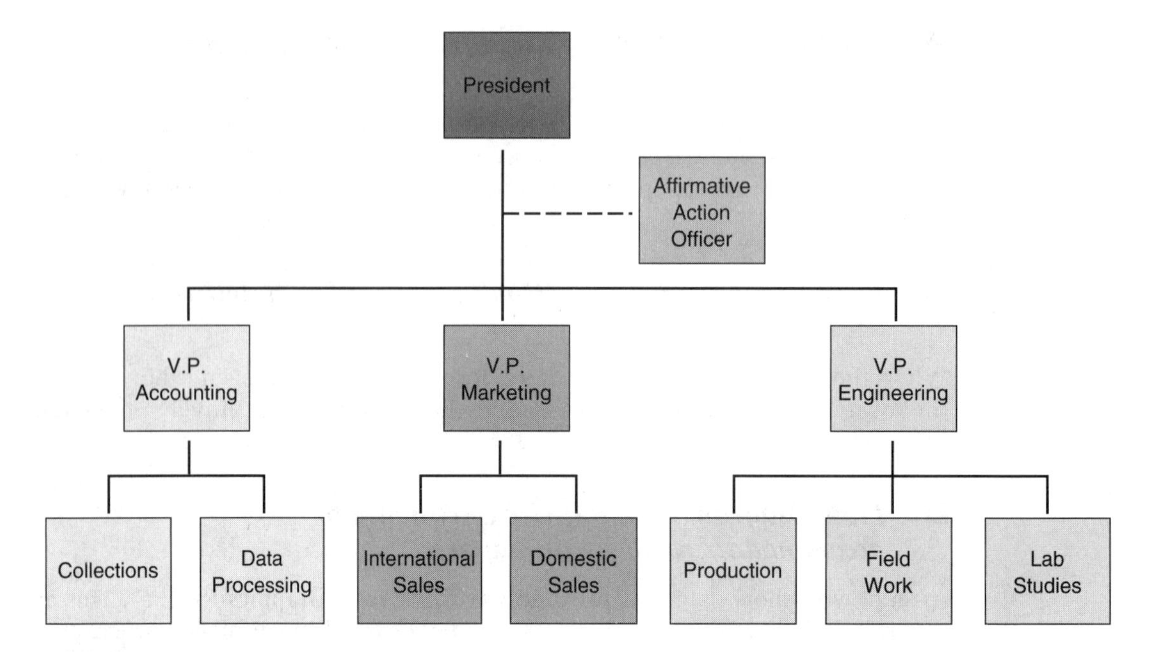

FIGURE 5–16
Basic organization chart

an example of this approach. In it, the president box is placed at the top; the vice president boxes are in the middle; and the individual department boxes are at the bottom. The size, border, and inside labeling of the shape can also show levels within an organization. For example, to show higher levels, the shapes can be made larger, have thicker border lines, or have labels in boldface. However, too many different shapes can clutter the chart, so vary shape size and appearance judiciously. Shading can also be used to highlight a portion of the chart. For example, the darker shading in Figure 5–16 could highlight the relationship between the president and the marketing organization.

■ *Organization Chart Rule #2: Connect shapes with solid or dotted lines.*

Use solid lines to show direct reporting relationships within an organization. Use dotted lines to show staff or indirect support relationships. In Figure 5–16, the affirmative action officer serves the president in a staff position, while the vice presidents serve in positions directly reporting to the president.

■ *Organization Chart Rule #3: Be creative to provide information and create interest.*

Use your imagination to create more informative and interesting organization charts. For example, consider the following ideas:

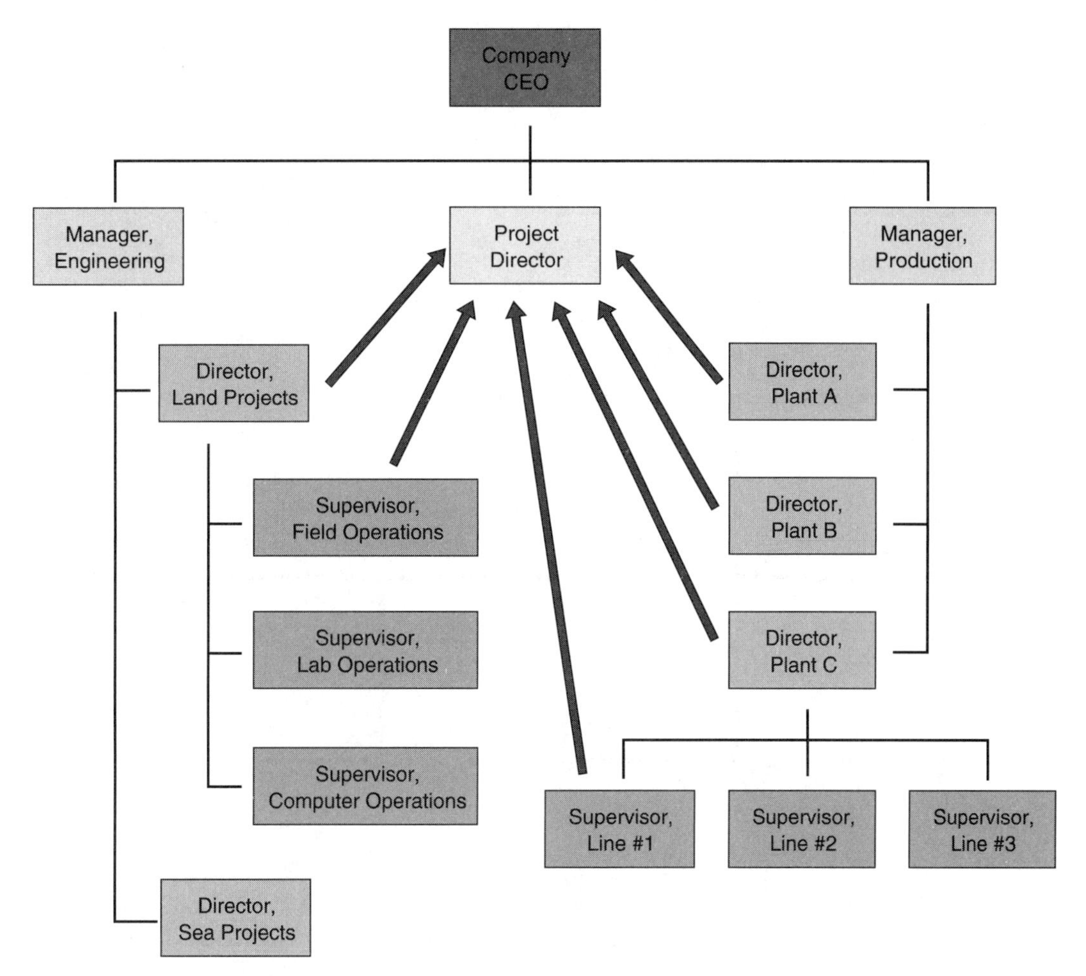

FIGURE 5–17
Adding to the basic organization chart

- Insert a photo of the assigned person within the applicable shape.
- Include a bulleted list of responsibilities inside or outside the shape for the applicable person, position, or work unit.
- Include shapes and lines to reflect your organization's interface with subcontractors, contract team members, and customers.

The following are examples of other creative approaches for organization charts:

- In Figure 5–17, sunlight-type rays emanate from the project director box to show the lines of responsibility in the project, yet the standard reporting lines remain.

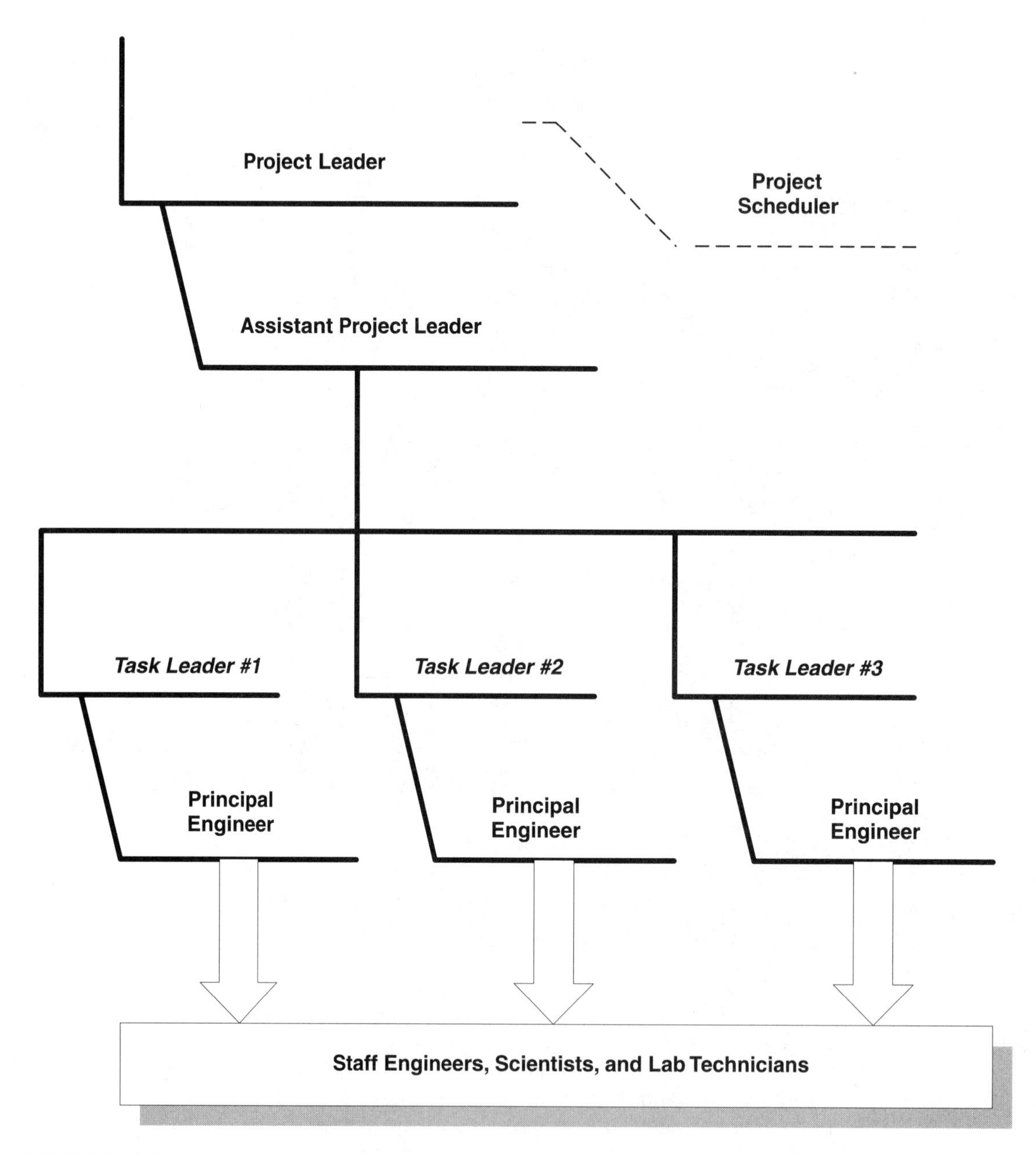

FIGURE 5–18
Line diagramming for an organization chart

- In Figure 5–18, organization levels are depicted with a line technique you may have used to diagram sentences in school.
- In Figure 5–19, concentric circles are used to show management levels within an organization.

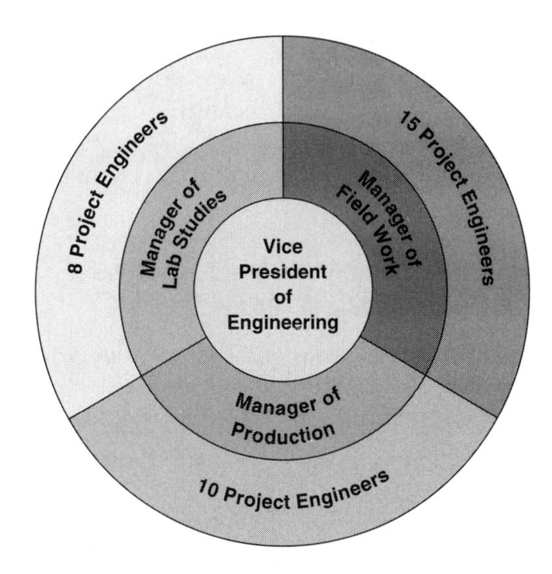

FIGURE 5–19
Concentric organization chart

Rules for Formal Tables

Tables present readers with information or raw data, in the form of either numbers or words. They are classified as either informal or formal:

- Informal—The table content is arranged in either rows or columns, with few if any headings. It has no table or graphic caption or identification number, is usually merged with text on a page, and isn't part of the proposal's figure or table list.

EXAMPLE

By the end of last month, we had shipped 22 new computers to our offices in Alabama. As shown below, most units went to Birmingham:

Birmingham	12
Montgomery	6
Mobile	4

- Formal—This table contains more detailed content arranged in a grid with horizontal rows and vertical columns that have headings. It is separated from the flow of the text, identified by a table or graphic number, and included in the proposal's figure or table list.

We recommend that only formal tables be used in a proposal. The exclusive use of formal tables will ensure that all tables have the same format and assigned captions and identification numbers. If you have information that would be appropriate for an informal table, present it in text or a bulleted list. The following rules and examples apply to formal tables.

■ *Formal Table Rule #1: Creatively use words and numbers to present data.*

There are many reasons and creative ways to present information and raw data in tables. Tabular formats can be used for the following purposes:

- Show responsiveness to requirements.
- List features, benefits, and other attributes.
- Show relationships among items.
- List procedural steps and activities.
- Summarize schedule information.
- Present numbers and totals.

Figure 5–20 shows an example for each of the preceding purposes. Each example uses a different table format. (If you use tables to show numerical totals, always verify the accuracy of these totals.)

■ *Formal Table Rule #2: Select the most appropriate format.*

Popular word processing programs have functions that allow you to build a customized table format or choose among a large selection of ready-to-use table formats. Your basic format decision will be how many vertical columns and horizontal rows will be needed to display the information or data. Other format decisions include how to frame the table and how to separate the columns and rows within the table. White, black, gray, or colored lines can be used as borders to frame the table. The following approaches can be used to define the column and row boundaries:

- White lines on a light gray background
- Light gray lines on a white background
- Alternating white and gray backgrounds
- Alternating colored backgrounds
- Horizontal and vertical lines in a grid pattern

If your backgrounds are too dark, they can make the table difficult to read. Backgrounds should therefore have gray screens no denser than 25 percent or be shaded with lighter colors.

Showing responsiveness to requirements:

Proposal Section	Applicable RFP Section Requirements
1. Experience	A.1-A.5
2. Management	A.6-A.7, B.1-B.4
3. Quality Control	C.1-C.3
4. Customer Service	C.4-C.5

Listing product features and benefits:

Features	*Benefits*
Customer support and repair facilities are within 20 miles of each of your 10 distribution centers.	• Reduces our technician travel time to serve your on-site needs, leading to a guaranteed one-hour response for on-site technical and repair assistance
Conveyor belt has proven reliability of 100,000 hours between failures.	• Lowers operational costs by reducing maintenance and out-of-service downtime of your distribution and sorting system

Listing features or attributes:

Name	Project Title	Yrs with Acme	Yrs in Computer Industry	College Degree
John Stern	Program Manager	15	20	PhD, Electrical Engineering
Bill Anders	Quality Manager	10	10	MS, Electrical Engineering
Mary Artman	Information Systems Manager	8	8	BS, Computer Science
Joe Vine	Customer Service Manager	9	15	BA, Business

Showing relationships among features:

	Required Aircrew Training Experience						
Instructor	Pilot	Navigator	Flight Engineer	Ordnance	Radar	Loadmaster	Acoustic Sensor
Mike Barnes	✔	✔					
Rod Stone	✔		✔				
Ron Strom		✔			✔		✔
Mark Vincent				✔		✔	

continued

FIGURE 5–20
Various formats and uses of formal tables

Listing sequential activities:

Step	Project Management Actions	Responsibility
1.	Report weekly status reports for each functional department.	All Department Mgrs.
2.	Review weekly status reports at department manager meeting.	Project Mgr.
3.	Compile and distribute monthly status reports to all functional departments.	Administrative Support Mgr.
4.	Compile and distribute quarterly status reports to customer.	Customer Service Mgr.

Listing scheduled activities:

Task	Responsible Department	Estimated Labor Hours	Completion Date
Feasibility Study	Engineering	75	1 March
Product Mock-up	Manufacturing	125	1 April
Development Testing	Test and Evaluation	50	15 April

Listing and compiling numbers:

		1999 Sales		
Region	1st Qtr	2nd Qtr	3rd Qtr	4th Qtr
U.S. North	2.56	3.10	3.80	4.23
U.S. East	1.89	1.58	2.14	3.08
U.S. South	1.57	2.53	2.89	3.37
U.S. West	2.46	2.89	3.19	3.48
International	4.12	5.89	5.94	6.23
Totals	$12.60	$15.99	$17.96	$20.39
Note: Dollars are shown in millions.				

FIGURE 5–20—*continued*
Various formats and uses of formal tables

■ *Formal Table Rule #3: Use style and format standards.*

As you develop your tables, follow the style and format standards listed in Table 5–1.

Rules for Drawings

Drawings can effectively depict physical objects, human or mechanical actions, and conceptual ideas. The following rules will help you create and use drawings.

TABLE 5–1: Style and Format Standards for Formal Tables

Topic	Standard
Headings	Create short, clear headings for all columns and rows.
Abbreviations	Include in the headings or the body of the table any necessary abbreviations or symbols. Spell out abbreviations and define terms in a key or footnote if you think the readers will need the clarification.
Numbers	For readability: • Round off numbers as much as possible • Align multidigit numbers on the right edge or at the decimal when a decimal is used.
Notes	Place explanatory headnotes either between the caption and the table—if the notes are short—or at the bottom of the table.
Footnotes	Place any necessary footnotes below the table.
Sources	Place any necessary source references below the footnotes.
Case	To enhance readability, don't use full caps in the main body of the table.
Clarity	Make each table self-contained and clear.

■ *Drawing Rule #1: Creatively display information.*

Used individually or combined with other types of graphics, drawings can be used to creatively display details in your proposal graphics. Keep the drawings as simple as possible, using only the detail that serves your purpose.

The following examples provide ideas for the creative use of drawings:

EXAMPLES

- Use geographic maps and facility/building layouts to provide location and spatial arrangement information. Within these drawings, add photographs, icons, and other drawings to add more detail.
- Within a system block diagram or schematic, insert drawings for the applicable components. In addition to drawings, photos can be used to depict components.
- To amplify a schedule, use drawings to illustrate a task or product represented by a scheduled milestone.
- To portray steps in a process flow diagram, place drawings of the action within the flow boxes or next to the connecting lines.

■ *Drawing Rule #2: Choose the most appropriate view.*

Drawings permit you to choose the level of detail needed. In addition, they offer you the following options for perspective or view:

- Exterior view—showing the surface features with a two- or three-dimensional appearance
- Cross-section view—showing a "slice" of the object, so that the interior can be viewed
- Exploded view—showing a relationship among parts by "exploding" or amplifying a portion of the object, as shown in Figure 5–21
- Cutaway view—showing the inner workings of object by removing part of the exterior

A photo is an effective way to describe an object and to show that a product exists. However, when a photo isn't available or doesn't show the needed detail, a drawing can be a good substitute.

■ *Drawing Rule #3: Clearly identify parts of the drawing.*

Place labels on every part of the drawing that you want your reader to see. (Conversely, you can choose *not* to label the parts that are irrelevant to your purpose.)

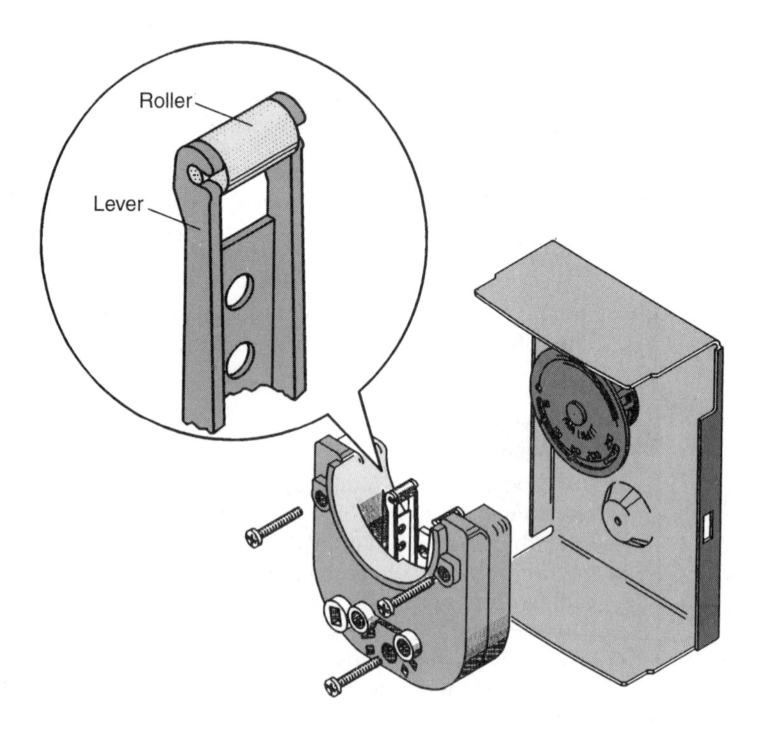

FIGURE 5–21
Exploded view in a drawing

Use a label typeface large enough for easy reading. Also, arrange labels so that (1) they are easy to locate and (2) they don't detract from the importance of the drawing itself. Avoid cluttering the graphic with too many labels. If a drawing has many parts that require identification, consider using number labels explained in a legend placed next to the drawing. Figure 5–22 shows the use of such a legend.

■ *Drawing Rule #4: Use clip art to supplement your own drawings.*

To supplement your own drawings, use clip art drawings from digital commercial or Internet sources. Ensure that the clip art is compatible with the context and tone of your proposal document. Indiscriminate use of general topic clip art can give the proposal a "comic book" look and make it appear you forced its use

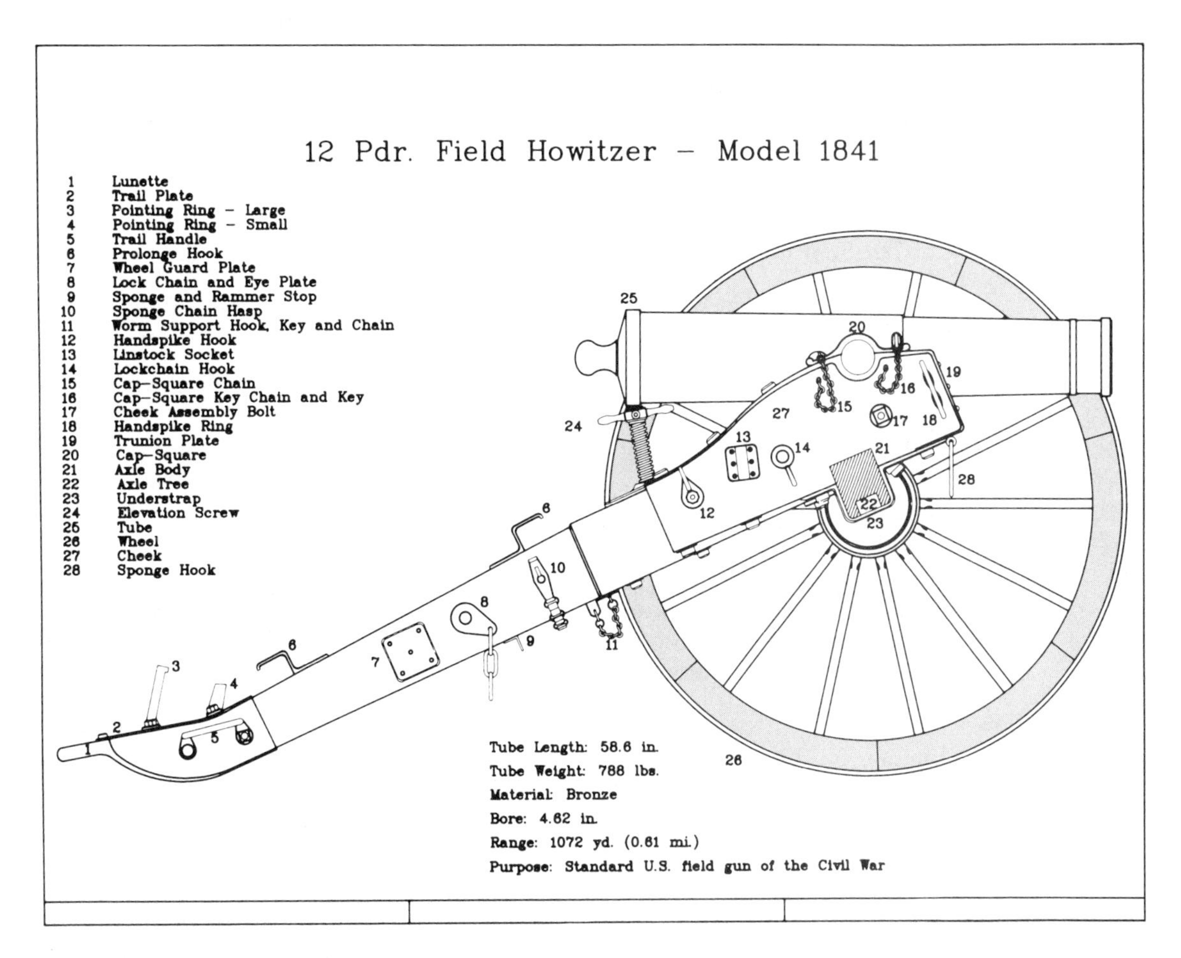

FIGURE 5–22
Using a legend in a drawing

simply because it was available. When you can, modify clip art to make it more applicable for your use.

Rules for Photos

Advances in digital photography and graphics software have made it easier and less expensive to use color and black-and-white photos in proposals. Use the following rules to select, produce, and display photos for your proposal.

■ *Photo Rule #1: Use photos to show that something exists.*

Use photos of your product or service to show that what you're offering is real—not a design or concept under development. Although a drawing of an existing product can be effective, its photo gives it a stronger impression of availability. Even if a product is in development, the photo of a conceptual model or a working prototype of the product can make it appear more than an idea. Of course, clearly identify what's in the photo; don't mislead readers into thinking that something is a proven product when it's only a conceptual model.

■ *Photo Rule #2: Crop photos to accentuate areas of interest and efficiently use proposal space.*

The trimming of unwanted portions of a photo is called cropping. Cropping provides the following benefits:

- It saves space. By cropping out unneeded parts of a photo, you don't waste proposal space with ineffective photo images.
- It helps to control the reader. Cropping around a certain part of a photo helps you direct readers to detail that you consider important. As you crop, also remember that within the image of the photo, the flow of the depicted action or the direction in which the eyes of the subject(s) are looking can also affect the movement and focus of your readers' eyes.
- It allows you to fit the photo into a desired page layout. By adjusting the overall size and the horizontal or vertical orientation of the photo, you can adjust the photo image to fit its assigned space in the layout.

Figure 5–23 shows how cropping can improve the effectiveness of a photo. Cropping can be performed by cutting the margins of photo prints, by adjusting the image of photos as they are processed in the darkroom, or, with graphics software, by electronically adjusting the image of digital graphics.

■ *Photo Rule #3: Use photo labels.*

Use labels (1) to describe action or identify key objects in your photos and (2) to support information in the text and the graphic's caption. The use of digital photos and graphics software allows you to insert labels electronically into your photos.

FIGURE 5–23
Photo cropping: before and after

Photo Rule #4: Use creative photo images and displays.

Photos offer many creative chances to improve the look and effectiveness of your proposal. For example, your photos can be modified by contrasting, tinting, and screening their image. Photos can also be laid out in a collage or combined with other types of graphics. Combining photos with other graphics can be especially effective. For example, a schematic of an electronic appliance could use lines for the electronic connections, photos for the major components, and drawings for minor subcomponents. Talk with illustrators and photographers to develop creative ideas.

Photo Rule #5: Plan ahead to identify and to meet your photo needs.

Collecting and processing photos can require substantial time. Therefore, identify your photo needs as early as possible. Use your storyboards to identify the photos that will be used in the main body of the proposal. Also, determine if photos will be used for the cover, tabs, title page, appendices, or other parts not in the main body. Allow enough time in your proposal preparation schedule to obtain and process your photos, and then to merge them into the proposal layout.

If you need the services of a photographer, ensure that the photographer clearly understands your needs. Specify the content of the photo and whether it's to be shot in color or black-and-white. (Avoid using color negatives or prints as the source for black-and-white prints; the images of the resulting black-and-white prints can have poor contrast and "washed out" detail.) In addition, indicate the planned size and orientation—horizontal, vertical, or square—of the photos in the proposal layout. The size can affect the choice of film speed, because photo images from high-speed film can become grainy and less detailed when enlarged. Orientation can dictate how the photographer will frame the image during the photo shoot.

Photos can also be digitized by scanning existing prints or shooting the photo with a digital camera. If you have a choice between scanning a photo in a printed document—for example, an old proposal—and scanning the original print, choose the original. The quality of a scanned original will probably be better. Digital photos can also be obtained from commercial clip art packages and free internet sources. Be careful to comply with all applicable copyright and usage rules for photos regardless of their source. Commercial software programs offer an array of features for adjusting the color, contrast, size, and other features of the digital photos.

> To justify its use, every graphic must make a valuable contribution to the proposal by presenting important information. Beware of fabricating a graphic solely because someone decides "We need more graphics" or of including a graphic because it "looks so good that we should use it somewhere." The most important aspect of a graphic is its content. A good graphic clearly and concisely tells an important story.
>
> (O'Connor, 1993, p. 238)

SUMMARY

Graphics (also called illustrations or visual aids) help make a proposal come alive for the reader. They come in the form of tables and figures, a term for all graphics other than tables. Use the graphics that best meet the needs of your readers. Whatever graphics you select, remember the four main reasons for using them: to simplify ideas, to reinforce ideas, to stimulate interest, and to support your proposal text. The graphics often used in proposals are pie charts, bar graphs, line graphs, schedules, flow diagrams, organization charts, tables, drawings, and photos.

Use the following general rules for graphics:

1. Follow a standard process for development and revision.
2. Follow style standards for consistency.
3. Use tracking and identification systems.
4. Refer to all graphics.
5. Place graphics near their first reference.
6. Strive to arrange graphics to be read without turning the proposal.
7. Design graphics for readability and comprehension.
8. Avoid clutter.
9. Provide support information.
10. Use color for a purpose.
11. Use typographical and design features as graphic components.

Use the following specific rules for nine common graphics.

Pie Chart Rules

1. Use no more than six or seven wedges.
2. Start clockwise at 12:00, moving from the largest to the smallest wedge.
3. Use a wedge scale divisible by 100.
4. Be creative but simple.
5. Draw and label carefully.

Bar Graph Rules

1. Use a limited number of bars.
2. Show comparisons quickly and accurately.
3. Vary bar spacing; maintain bar width.
4. Arrange bars in an order that best suits your purpose.
5. Creatively show comparisons.

Line Graph Rules

1. Use line graphs to stress trends.
2. Place line graphs strategically.
3. Strive for accuracy and clarity.
4. Avoid including numbers on the line graph itself.
5. Use multiple lines with care.

Schedule Rules

1. Show only main activities and use an appropriate time scale.
2. List activities in sequence, starting at the top.
3. Run all labels in the same direction.
4. Creatively develop different formats.
5. Be realistic in your schedule.

Flow Diagram Rules

1. Present only overviews.
2. Limit the number of shapes.
3. Provide a legend that defines the shapes.
4. Use a sequence to accommodate reader scan patterns.
5. Label all shapes.

Organization Chart Rules

1. Use shape placement and features to show organization levels.
2. Connect shapes with solid or dotted lines.
3. Be creative to provide information and attract interest.

Formal Table Rules

1. Creatively use words and numbers to present data.
2. Select the most appropriate format.
3. Use style and format standards.

Drawing Rules

1. Creatively display information.
2. Choose the most appropriate view.
3. Clearly identify parts of the drawing.

Photo Rules

1. Use photos to show that something exists.
2. Crop photos to accentuate areas of interest and efficiently use proposal space.
3. Use photo labels.
4. Use creative photo images and displays.
5. Plan ahead to identify and to meet your photo needs.

EXERCISES

1. Graphics Conceptualization

Look at the formal proposal model in Chapter 7. Suggest ideas for graphics that weren't used in the example, providing one suggestion for each of the nine types of graphics described in this chapter.

2. Graphics Critique

Go through periodicals, such as newspapers and magazines, and find two examples of each type of the nine types of graphics described in this chapter. Critique the effectiveness of each graphic based on its layout and content.

3. Pie Chart, Bar Graph, and Line Graph

Complete this exercise using the data in exercise Table 5–2. (This table shows total U.S. energy production and consumption from 1960 to 1985. It also breaks

TABLE 5–2: Energy Production and Consumption, by Major Source: 1960 to 1985

[Btu = British thermal unit. For Btu conversion factors, see text, section 19. See also *Historical Statistics, Colonial Times to 1970,* series M 76–92]

Year	Total production (quad. Btu)	Percent of Production				Total consumption (quad. Btu)	Percent of Consumption				Consumption/production ratio
		Coal	Petroleum[1]	Natural gas[2]	Other[3]		Coal	Petroleum[1]	Natural gas[2]	Other[3]	
1960	41.5	26.1	36.0	34.0	3.9	43.8	22.5	45.5	28.3	3.8	1.06
1961	42.0	24.9	36.2	34.9	4.0	44.5	21.6	45.5	29.1	3.8	1.06
1962	43.6	25.0	35.6	35.1	4.2	46.5	21.3	45.2	29.5	4.0	1.07
1963	45.9	25.8	34.8	35.4	4.0	48.3	21.5	44.9	29.8	3.7	1.05
1964	47.7	26.2	33.9	35.8	4.0	50.5	21.7	44.2	30.3	3.8	1.06
1965	49.3	26.5	33.5	35.8	4.3	52.7	22.0	44.1	29.9	4.0	1.07
1966	52.2	25.8	33.7	36.4	4.1	55.7	21.8	43.8	30.5	3.8	1.07
1967	55.0	25.1	33.9	36.5	4.4	57.6	20.7	43.9	31.2	4.2	1.05
1968	56.8	24.0	34.0	37.6	4.4	61.0	20.2	44.2	31.5	4.1	1.07
1969	59.1	23.5	33.1	38.7	4.8	64.2	19.3	44.1	32.2	4.4	1.09
1970	62.1	23.5	32.9	38.9	4.7	66.4	18.5	44.4	32.8	4.3	1.07
1971	61.3	21.5	32.7	40.5	5.3	67.9	17.1	45.0	33.1	4.8	1.11
1972	62.4	22.6	32.1	39.7	5.6	71.3	16.9	46.2	31.9	5.0	1.14
1973	62.1	22.5	31.4	39.9	6.2	74.3	17.5	46.9	30.3	5.3	1.20
1974	60.8	23.1	30.5	38.9	7.4	72.5	17.5	46.1	30.0	6.5	1.19
1975	59.9	25.0	29.6	36.8	8.6	70.5	17.9	46.4	28.3	7.4	1.18
1976	59.9	26.1	28.8	36.4	8.6	74.4	18.3	47.3	27.4	7.1	1.24
1977	60.2	26.2	29.0	36.4	8.5	76.3	18.2	48.7	26.1	7.0	1.27
1978	61.1	24.4	30.2	35.6	9.9	78.1	17.6	48.6	25.6	8.1	1.28
1979	63.8	27.5	28.4	35.0	9.1	78.9	19.1	47.1	26.2	7.7	1.24
1980	64.8	28.7	28.2	34.2	8.9	76.0	20.3	45.0	26.8	7.8	1.17
1981	64.4	28.5	28.2	34.2	9.1	74.0	21.5	43.2	26.9	8.4	1.15
1982	63.9	29.2	28.7	32.0	10.2	70.8	21.6	42.7	26.1	9.6	1.11
1983	61.2	28.2	30.1	30.6	11.2	70.5	22.6	42.6	24.6	10.2	1.15
1984	65.9	30.0	28.6	30.7	10.7	74.1	23.0	41.9	25.0	10.1	1.13
1985, prel	64.7	30.0	29.2	29.6	11.3	73.8	23.7	41.8	24.1	10.5	1.14

[1]Production includes crude oil and lease condensate. Consumption includes domestically produced crude oil, natural gas liquids, and lease condensate, plus imported crude oil and products. [2]Production includes natural gas liquids; consumption excludes natural gas liquids. [3]Comprised of hydropower, nuclear power, geothermal energy and other.
Source: U.S. Energy Information Administration, *Annual Energy Review.*

down production and consumption into four main groupings.) Graphics for this exercise can be drawn by hand or a software program.

a. Construct a pie chart that reflects the four groupings of percentage breakdowns for U.S. energy consumption in 1980.
b. Construct a grouped bar graph (see Bar Graph Rule #5) that reflects both U.S. total production and U.S. total consumption figures for 1960, 1965, 1970, 1975, 1980, and 1985.
c. Construct a line graph that reflects the trend in U.S. total energy consumption from 1970 through 1979.

4. *Schedules*

Create the following schedules:

a. Construct a horizontal bar schedule (Gantt) to reflect your work on a past, present, or future writing project.
b. Select a project that you plan to complete on the job or in a course. Construct a milestone schedule that shows some or all of the planned activities.

5. *Flow Diagrams*

Construct a flow diagram that describes a process with which you are familiar. For example, you may want to base the flow diagram on the activities reflected in the schedules made for Exercise #4.

6. *Organization Charts*

Construct the following organization charts:

a. Create an organization chart with concentric circles. Use personnel data from your own work experience, from interviews with members of other organizations, or from library research about companies.
b. Create a conventional organization chart with geometric shapes. Use personnel and organizational data based on your involvement in social, professional, school, or employment settings.

7. *Table and Figure from Same Data*

Construct a formal table that includes all data in the following fictional case; then create another type of figure that highlights some of the data in your table.

Design the table to show the significant sales increases in six main food products of your firm, Tasty Foods, Inc., and to reflect the increasing percentages of product sales being exported. The six products are diverse, so some kind of grouping will be appropriate in the table. Use the following information to construct the table and figure. (Note: all dollar and percentages are annual totals for 1997, 1998, and 1999, respectively; all dollar figures are given in millions.)

- The annual sales of your best-selling product, Basetone Beer, were $23, $34, and $43 from 1997–1999.

- Another big seller was Castle Cake Mix, with annual sales figures of $18, $19, and $21 during the three years.
- Both Basetone Beer and Castle Cake Mix had the same export percentages of sales: 12 percent, 18 percent, and 21 percent during the three years.
- Your Brandy's Brownies, a relatively new entry, had $2, $4.2, and $6 in annual sales for the three years, with annual export percentages starting at 34 percent in 1997 and then increasing to 38 percent and 41 percent in the succeeding years.
- Another new entry, OK Orange Drink, is having a bit harder time in its competitive market: over the three years, annual sales were $1.2, $1.3, and $1.35, with annual export percentages of 3 percent, 6 percent, and 4 percent, respectively.
- Finally, two old reliable products, Gramp's Granola Bars and Tangy Tea, are holding their own. The first product went from $21, to $14, and then to $15 in annual sales during the three years, with annual export percentages of 32 percent, 42 percent, and 42 percent, respectively. Tangy Tea had lower annual sales of $3, $4.5, and $6.1 during the three years, but its annual percentages were 45 percent, 59 percent, and 71 percent, during the period.

8. Formal Table

From any chapter in this book, develop a formal table that shows at least five sets of features and benefits that reflect rules, tips, or other recommendations in that chapter.

9. Drawings

Find drawing examples from periodicals, such as newspapers and magazines, to show one example of each of the four perspectives described in Drawing Rule #2 (choose the most appropriate view). From these periodicals, also show any other examples of creative drawings.

10. Photos

Find photo examples from periodicals, such as newspapers and magazines, to show photos that would be more effective with cropping or labels.

Sales Letters and the Executive Summary

OBJECTIVES

- **Learn the reasons for writing sales letters**
- **Learn the rules for writing sales letters**
- **Examine model sales letters**
- **Learn the reasons and rules for developing executive summaries**
- **Perform exercises using chapter guidelines**

The sales letter and executive summary are used to support the proposal. They can greatly affect the impact of your proposal on the customer, and, in the case of a sales letter, can determine if you'll even have the chance to submit a proposal. The content of sales letters and executive summaries varies.

In this chapter, we'll look first at writing sales letters. Chapter 1 stressed the importance of contacting the prospective customer many times, both in person and in writing. As one method of contact, the sales letter gives you the opportunity to influence your customer. The executive summary, submitted with the proposal, can also influence your customer. The latter portion of this chapter provides rules for planning and developing the executive summary.

WHY AND WHEN TO WRITE SALES LETTERS

> Letters are one of the principal forms of business communication. Good letters do more than convey information, actions, and decisions. They establish the personal style of the sender and the image of the sender's organization, and they act as surrogate conversations between parties.
>
> (Freeman and Bacon, 1990, reprinted 1995, p. 112)

TABLE 6–1: Uses of Sales Letters

Reason	Content Example
Start a relationship	"I'll be calling you on Friday to talk about how we might be of service to your branch offices."
Follow up a phone call	"It was good to talk with you. Can we meet next week to discuss your needs regarding. . . ?"
Follow up a meeting	"You mentioned that you could use more information on . . . , so here's an article that might interest you."
Accompany the delivery of a formal proposal	"The major benefit of this proposal is the expected reduction of your accounting costs by 50 percent over the next year."
Follow up after your proposal is announced as a winner	"You have chosen the firm that will pay close attention to your need for quality work. You have my personal commitment that we will meet all contract requirements."
Follow up after your proposal is announced as a loser	"We learned a great deal about your firm and hope to do business with you another time. Was there some way our proposal could have been made more responsive to your needs?"
Cushion the impact of an invoice	"I have appreciated your business. I will call next week to see if you have any further follow-up questions."
Seek repeat business	"We enjoyed working with you last October. To resume our business relationship, we now offer a new product that might interest you."

You may think that your sales proposal is a stand-alone document that succeeds or fails on its own merits. However, its success often depends on the customer contacts you make before and after proposal delivery. The following are three main reasons for making sales letters an important part of your proposal process.

■ *Reason #1: To reinforce*

Letters help reinforce and clarify what is discussed in person. You can use them to summarize a discussion, provide a formal record of points discussed, and highlight important issues.

■ *Reason #2: To follow up*

Letters give you the opportunity to respond to questions you weren't able to answer during a meeting. The customer can be favorably impressed by your follow-up effort to resolve these questions. You can also use a letter to send additional material, such as brochures, specifications, flyers, and articles, that you think will help the customer.

■ *Reason #3: To exercise control*

Letters help you to maintain control of the proposal process and to remind customers that you're still there, ready to do whatever is necessary to meet their needs.

Letters can be sent to the customer during all stages of the sales proposal process described in Chapter 1. Table 6–1 lists some specific reasons—before and after proposal submittal—for sending sales letters.

RULES FOR WRITING SALES LETTERS

Some people seem to have a flair for writing, just as some people seem to be "born" salespeople. Writing good letters, however, is much more a matter of applying basic communication strategies than it is an innate gift. You can develop competence in letter writing, as in sales, by learning some fundamentals and then putting them into action.

The key is to discover just what the reader wants from the letter. Respond to a reader's needs by writing a visually appealing letter and by making important information easy to find. Use the following guidelines to write sales letters that satisfy the needs of your customer.

■ *Letter Rule #1: Use the "easy-in, easy-out" rhetorical pattern.*

Readers are most likely to give attention to "easy-in, easy-out" letters—that is, those that are easy to begin reading and easy to exit. Figure 6–1 shows a visual pattern for sales letters that responds to this reader need. Short first and last paragraphs quickly give readers the impression that they can read your letter in a minute or less. If the format immediately creates this feeling of comfort, your message can attract the reader's attention.

■ *Letter Rule #2: Use the 3-Cs pattern to gain and keep reader interest.*

You can gain and keep the reader's interest by following this simple three-part guideline for the 3 Cs:

1. *C*apture interest with a good opener.
2. *C*onvince with powerful supporting points.
3. *C*ontrol response by making clear the next step.

Even if readers have already expressed interest in what you're selling, don't assume that they'll anxiously be waiting for your letter. Because reading any document can

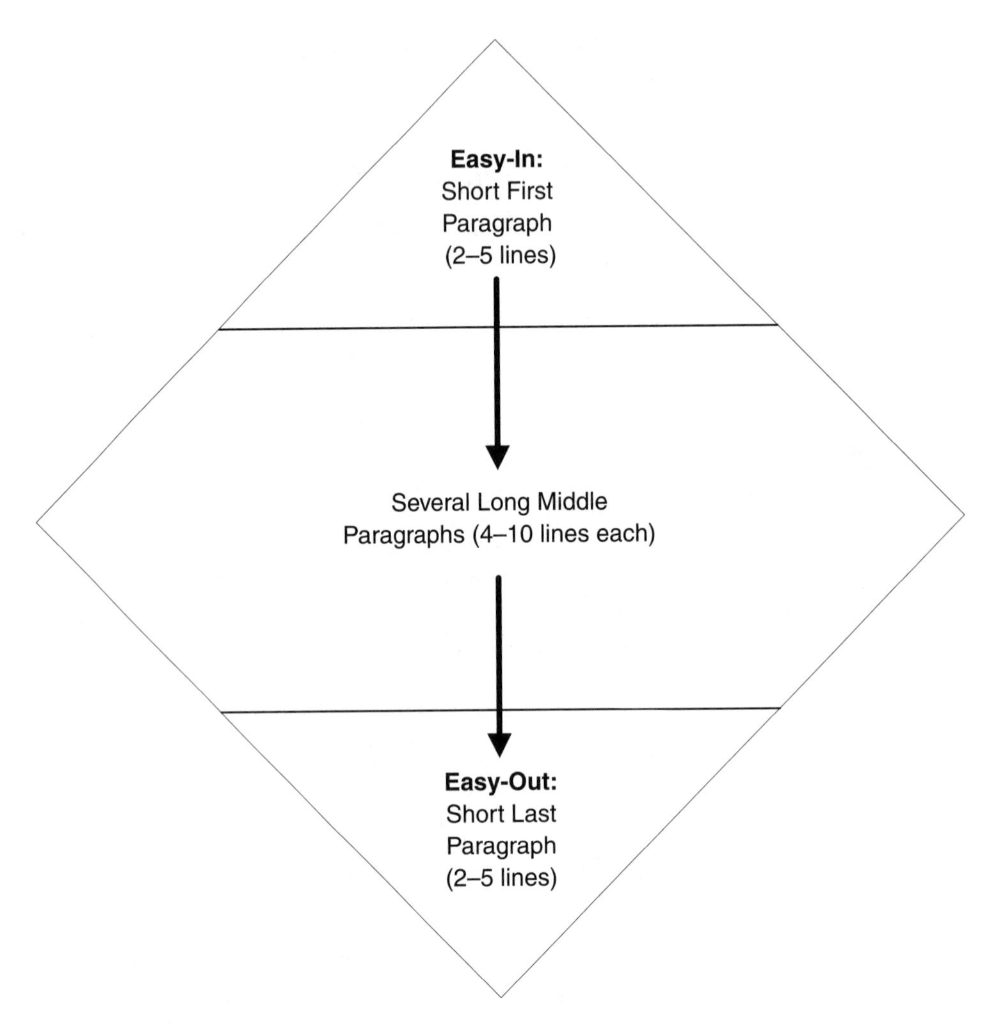

FIGURE 6–1
Visual image of the "Easy-in, Easy-out" approach

be a chore for busy people, use the 3-Cs pattern to entice readers to start and finish the reading of your letter. The following three rules support the use of this pattern.

■ *Letter Rule #3: Capture interest with a strong first paragraph.*

Readers can be most receptive at the start of a letter when they want to know who you are and how your letter can help them. Capture their attention in the first paragraph by (1) showing that what you have to say will help solve a problem and (2) providing a reference or link to your last contact with the reader.

For example, use the following techniques to attract the reader's interest in the first paragraph:

- Cite a surprising fact.
- Announce a new product or service.
- Ask a question.
- Show understanding of the customer's problem.
- Show potential for solving the customer's problem.
- Present a supportive testimonial about your product or service.
- Make a challenging claim about your product or service that can be substantiated.
- Summarize the results of a meeting with the customer.
- Present a "what if" scenario.
- Answer a question that the reader previously asked.

Let's look at a good and bad example of an opening paragraph:

EXAMPLE

- I enjoyed meeting with you yesterday and learning about your product lines. You noted that lathe #7 has a bearing problem. Have you considered solving the problem with a change out? Here's how it might work.

  CRITIQUE: This paragraph uses natural language and an unobtrusive reference to a meeting held the previous day. Most important, it focuses on the customer's problem and then suggests a way to solve it. The question ("Have you considered . . . ?") encourages the reader to continue.

EXAMPLE

- It is our pleasure to enclose a recent brochure pursuant to your request for more information about our new-generation ball bearing product. We believe this product—the pride of our firm—offers state-of-the-art technology in accordance with stringent industry standards and specifications.

  CRITIQUE: This paragraph starts with a superficial opening ("It is our pleasure . . .") and uses dry jargon ("pursuant to" and "in accordance with"). It also focuses almost exclusively on the writer's view of the offered product, rather than the reader's concern about the bearing problem. Write the sales letter without trivial, stilted phrases and overt bragging, and approach the problem from the reader's perspective, even to the point of using the pronoun *you* rather than *I* and *we*.

Letter Rule #4: Convince readers in the body paragraphs.

After gaining your readers' attention, follow through on whatever promise the first paragraph held. Following are some guidelines for writing the body paragraphs:

- See every point from the readers' perspective. Word every sentence to attract their interest and keep them reading past the first paragraph.
- Use a deductive (general-to-specific) plan of organization in each paragraph. Your first sentence, often the topic sentence, should state the point most important to readers. The remaining sentences should support the first one. Unfortunately, writers often bury main points in the middle of paragraphs or lead up to them at the end. Failing to include the main point in the first sentence assumes that readers have time to discover your purpose. They may not. If a reader doesn't find an important or interesting idea early in the letter, they may lose interest and not read the rest of the letter.
- Use the customer's name in the body of the letter. Leading a paragraph with a sentence such as "Ms. Janowsky, you mentioned that you wanted to modernize your filing system" gives the personal touch that can help build relationships. Whether you use the first or last name depends on how you address the reader in person.
- Limit the selling points you stress in the letter. Avoid the shotgun approach, in which you blast the readers with a long list of your company's attributes. Instead, focus on about three selling points (themes) that you think will be most appealing to your readers. For example, your selling points could be your firm's quality work, service after the sale, or leadership in research and development.
- Transition from what readers need to what you have to offer. Once you have established need, there's a better chance that readers will want to learn more about what you have to offer. Emphasize what's unique about your firm or idea. Your best selling points can be those that positively differentiate you from your competitors.
- Explain the value of enclosures. Items such as brochures and specification sheets won't necessarily get the attention they deserve if you don't mention their value. Enclosures should be "sold," not just sent. Also, if your letter gets separated from the enclosures, a letter that's clear about the importance of the enclosures can encourage the reader to find and read them.
- Link value with price, if price needs to be mentioned at all. You don't have to be defensive about the cost of your product or service, if the customer understands its value. Also, don't put a lot of dollar figures in paragraph form, for they can be hard to find and therefore tend to make readers suspicious of your prices. Instead, put prices in lists or tables, either within the letter or in an attachment.

■ *Letter Rule #5: Control the next move with the last paragraph.*

Write the last paragraph to leave yourself, not the reader, in control of the next step. You could conclude with comments such as these:

- "I'll give you a call within a few days to arrange an appointment time that's convenient for you."
- "The proposal will be in the mail next week."
- "Bob and I are looking forward to our meeting scheduled for next week."

Don't assume that your customers will call you after they receive your letter. Even if they call, you may have little control over when they do. Thus, the referral to a follow-up call can be an excellent way to end a letter. Keep a reminder file, so that you actually make those promised calls.

■ *Letter Rule #6: Break up body paragraphs for easy reading.*

The following are a few structural devices for organizing your letter into discrete "chunks" of information:

- Numbered points
- Bulleted points (such as those in this list)
- Subheadings

When using any of these devices, try to develop points in groups of three. Such groupings create a rhythm, attract attention, and encourage recall. Also, a short listing is more appropriate for a short document with a brief message, such as a letter.

■ *Letter Rule #7: Add a postscript for special attention.*

A good letter has the first paragraph capture interest, the middle paragraphs convince the reader, and the last paragraph control the outcome. But you might have something else to say—a special offer, a reminder about a meeting, or reinforcement of a main point already mentioned—that you want to stand out in the letter. For these purposes, a P.S. may be appropriate. Readers may actually pay more attention to the P.S. than to the main part of the letter. In addition, a handwritten P.S. on a typewritten letter can make the letter appear more personal, in addition to drawing attention to points of significance.

However, some readers may not consider the P.S. an appropriate appendage to a formal business letter. Readers may also think that a P.S. reflects the writer's inability to organize the content of the letter. So in deciding whether to use a P.S., judge the background and expectations of the readers.

■ *Letter Rule #8: Write one-page letters, using attachments for details.*

There are two good reasons to write one-page letters: (1) many people in business prefer the one-page letter and (2) letters should be short because they should complement personal contacts with clients—not replace them. When you have a lot to say, use attachments to shorten the main body of the letter. Attachments can also help you separate main points in the letter from less important supporting details. For example, a letter suggesting two alternatives to solve an accounting problem could highlight the alternatives in the letter and more fully describe the alternatives in an attachment or two.

Sometimes a letter must run longer than a page. For example, you may find that to answer a series of questions posed by the reader will take more than one page. To keep the reader's interest in a multipage letter, end pages with a comment or rhetorical question that encourages the reader to turn to the next page. For example, a question such as "Just how much will you save with this new inventory technique?" might come right before a page change.

■ *Letter Rule #9: Use the letter style your readers prefer.*

You may dismiss the selection of letter styles (block, modified block, or simplified) as inconsequential, and accept the style most familiar to your secretaries or firm. But this approach misses an opportunity to show customers you're ready to meet their standards. Instead, match your letter style to theirs. Differences among the main letter styles are largely mechanical and can be mastered quickly. Examples of sales letters with different styles and purposes are shown later in this chapter.

■ *Letter Rule #10: Edit carefully.*

Edit your letters carefully to ensure that there are no style, grammatical, or mechanical errors. These kinds of errors in a short document can lead readers to think: if you can't produce an error-free letter, how can they expect you to provide a quality product or service? Chapter 10 provides more details about editing.

TIPS FOR WRITING TRANSMITTAL LETTERS

> The cover letter is part of the proposal package and should help sell your solution. It should reference the RFP by name or number and indicate the effective period for the prices and implementation schedules outlined in the proposal. It can also mention the key selling benefit of your proposal. The cover letter is also the place to thank individuals in the client's organization who have been helpful to you. . . . If your proposal is a revision, indicate that in the cover letter.
>
> (Sant, 1992, p. 118)

A transmittal letter—also known as a cover letter—is written to accompany the proposal document. It can be part of, or detached from, the proposal document. Its main purpose is to identify the proposal, including its source and intended recipient. However, a well-written transmittal letter can support your proposal by (1) summarizing background information and your key selling points, (2) indicating a strong commitment to back up the proposal, and (3) identifying the person to contact if there are any questions about the proposal. Of course, if the customer provides specific requirements about the content and format of the letter, and how it is to be sent with the proposal, comply with these instructions.

The following guidelines are important for writing a transmittal letter:

- Use short beginning and ending paragraphs—about three to five lines each.
- Use a conversational style, with little or no technical jargon. Avoid stuffy phrases such as "per your request" or "as contained herewith."

- Avoid trite and stale openings, such as "we are pleased to submit" or "we welcome the opportunity to propose." These types of lead-ins add little to your sales message and state the obvious.
- Use the first paragraph for introductory information, mentioning what your proposal responds to—for example, a formal RFP, a conversation with the customer, or your perception of a need. Whether it's here or elsewhere in the letter, ensure that you identify the applicable RFP and the title and if any, the number of your proposal.
- Use the middle of the letter to emphasize the key benefits of your proposal. Stress what you can do to solve a problem, using the words *you* and *your,* rather than *I, we,* and *our.* You can also state your strong desire and commitment to provide the proposed product or service.
- Use the last paragraph to retain control by orchestrating the next step in the proposal process. Indicate who the customer should contact for more information. When appropriate, indicate that you'll call the customer to follow up after proposal delivery.

EXAMPLES OF EFFECTIVE SALES LETTERS

This section provides models of five effective sales letters. To show how the letters accomplish their purpose, we describe each one in terms of the letter's style (block, modified block, or simplified), the letter itself (with marginal remarks), and a critique. *(Note: all of these model letters are fictional.)* The examples include the following types of letters:

- Letter of first contact, also known as a cold call letter—with block style
- Letter after a phone call—with modified block style, no paragraph indention
- Letter after a meeting—with modified block style, paragraph indention
- Letter of reacquaintance, also known as a refresher letter—with simplified style
- Letter of transmittal, also known as a cover letter—with block style

Example #1: Letter of First Contact (Cold Call Letter)

This letter, shown in Figure 6–2, is written with a block style in which the following letter parts begin at the left margin: date line, inside address, salutation, letter paragraphs, closing, name/title or department, and initials. Left justification promotes readability.

Sending your first letter to a potential customer is like sending a job letter and résumé: the only purpose is to get your foot in the door. Peterson wants to interest Taft just enough to get the phone conversation going next week, thus paving the way for a first meeting.

The following is a critique of this letter:

- *Paragraph #1*—Peterson will probably capture Taft's interest, for he immediately introduces a word of great concern to bankers: risk. Also, he uses the question pattern, which encourages the readers to read on to find the answer. Finally, he briefly mentions the solution—an environmental survey—that will

Uses a block style with major letter parts starting at the left margin (except the address on the letterhead)

Engages the reader by asking a question and introducing the subject of great interest: risk

Uses a statistic to show the extent of the problem

Shows that there's a solution

Uses bullets to highlight benefits of the environmental survey

Stays in control by noting that he'll call soon

Jones Engineering, Inc.

Geotechnical Consultants
213 Spring Street
Hardyville, Texas 77766

March 4, 2000

Mr. Thomas Taft, President
Faust Savings and Loan
23 Forge Avenue
Strutney, Texas 77112

Dear Mr. Taft:

Did you know that when you make a loan on industrial property where hazardous waste is disposed, your risk may exceed property value? Fortunately, there's a way to predict your risk: an environmental survey.

For the past 40 years, Gulf Coast plants have generated over 25 percent of the nation's hazardous waste. Most of that material is buried or dumped at sites near its origin. If your bank unknowingly holds a lien on a location where such wastes have been deposited, you may also hold all liability in the event of foreclosure.

Conducting an environmental survey before you make a loan greatly reduces your risk. One of our surveys, which can usually be completed in less than two weeks, will

- Identify potential waste problems
- Describe the scope of the needed clean-up
- Develop a cost estimate for clean-up

Mr. Taft, I'll give you a call next week to schedule an appointment for explaining how our site surveys can help Faust Savings and Loan reduce its liability risk.

Sincerely,

G. L. Peterson

Gannly L. Peterson, P.E.
Hazardous Waste Department

dv

FIGURE 6–2
Letter of first contact—block style

be described in the letter. In just two sentences, he has captured interest and has projected the points to come.

- *Paragraph #2*—The two middle paragraphs persuade by elaborating on the problem and suggesting a solution. Taft learns that his savings and loan makes commercial loans in a state with a massive hazardous waste problem. Then he discovers that foreclosure, usually considered the main way to reduce losses, can actually increase them.
- *Paragraph #3*—It's Jones Engineering to the rescue! Peterson uses three bulleted points to indicate how an environmental survey can help lower the liability risk of Faust Savings and Loan. He also includes the selling point that these surveys usually take less than two weeks, because he knows that Taft will not want the survey process to extend closings on commercial properties.
- *Paragraph #4*—Peterson remains in control of the sales process by mentioning that he'll call the next week for an appointment. His language is important: note that he states the purpose from the customer's point of view, not his own. He wants to explain "how our site surveys can help Faust Savings and Loan reduce its liability risk." Peterson wants to appear as a helper, not a pushy salesman.

Example #2: Letter After a Phone Call

This letter, shown in Figure 6–3, is written in a version of the modified block style. It differs from the block style in that three parts—dateline, closing, and name/title or department—are indented, much as they might be in an informal letter to a friend. All three are indented the same number of spaces, starting about two-thirds across the page. The paragraphs are not indented.

This kind of letter should be sent immediately after a first phone call with a potential customer. With Reuben having the customer's attention, if he demonstrates responsiveness to her needs, he'll be on the way to writing a proposal with a good chance of winning a contract. This letter, then, can provide the foundation for a productive meeting.

The following is a critique of this letter:

- *Paragraph #1*—The writer first mentions his previous contact with the reader, and then uses a lead-off technique to capture interest: a challenging claim that he can substantiate. For example, his proof for the claim "My research shows . . ." might be based on the postseminar evaluations made after teaching a seminar to another firm's seminar. This substantiation can be presented to Letson later in the proposal process, if necessary.
- *Paragraph #2*—It's important not to appear pushy in a letter that comes between a first phone call and a first meeting. From what Reuben says in the letter, it appears that Letson needs more time to ponder the course and to discuss it with her training staff. She might want to resolve some internal disagreements about the course before Reuben can be brought into the picture. Thus, enclosing the three relevant documents—tentative program outline, journal article, and résumé—is a good idea; everyone can look them over.

Uses a modified block style with indented date, closing, and name/title sections

Refers to previous contact (phone call) and engages interest with significant claim

Responds to reader's specific request

Uses bullets to emphasize three enclosures

Mentions topics that'll be covered in a future talk to produce a tailored training session

Stays in control by saying he'll call

Write Right! Inc.
Harold S. Reuben, Ph.D.
123 Upp Street
Baltimore, Maryland 22233
(212) 222-6622

May 15, 2000

Ms. Koro Y. Letson
Bradley Metal Products
202 Jetson Blvd.
Sunnyland, Maryland 21213

Dear Ms. Letson:

I enjoyed talking with you yesterday about your interest in a sales writing seminar. My research shows that this kind of seminar can increase your sales from 5 to 10 percent within a year.

You mentioned that you wanted some background information before we proceed to the proposal stage, so I have enclosed the following three items:

- A draft program outline for the seminar, based on the needs you described
- A recent article in *Sales Journal* about the value of a consistent company approach to sales proposals
- My résumé, including a list of recent clients

Let me stress that the enclosed program outline is just one possible approach. We can refine it more after we talk further about the goals of the seminar, the background of the participants, and your scheduling needs.

As you said, your training staff might want to help with seminar design. I'll give you a call next week to determine if we can meet to discuss your seminar needs.

Sincerely,

Harold S. Reuben

Harold S. Reuben
President

rf
Enclosures (3)

FIGURE 6–3
Letter after a phone call—modified block style, no paragraph indention

- *Paragraph #3*—Here Reuben mentions some specific types of information he'll need to plan a course, namely, data about the course goals, participants' backgrounds, and Letson's scheduling needs. These three topics may form an outline for a future conversation.
- *Paragraph #4*—Reuben remains in control of the proposal process with the promise to call next week. He gives Letson a week because he knows she needs time to consider the project before proceeding. A premature call might alienate her.

Example #3: Letter After a Meeting

This letter, shown in Figure 6–4, is also written in a modified block style. Its dateline, closing, and name/title or department items are indented at least halfway across the page. However, this version of the modified block style indents the first line of each paragraph five spaces.

It appears that Orley has the good fortune of helping the customer design the RFP. This opportunity may help Orley to write a winning proposal because (1) customers may feel inclined to hire the company that helped them determine what they want and (2) a company on this kind of inside track would probably suggest RFP requirements to which its firm would be best able to respond.

The following is a critique of this letter:

- *Paragraph #1*—Orley hasn't finalized this deal and has the sense not to appear overconfident. Instead, she alludes to the progress made at the meeting, a remark that helps cement her relationship with the customers. They may be starting to view Best as a member of their "team." Then Orley captures interest by projecting the content of the letter: it will summarize the results of the meeting and answer a question that Seuss posed.
- *Paragraph #2*—Orley writes convincingly here. By summarizing the main conclusions of the meeting, she reinforces the department stores' inclination to adopt a major service awards program. Orley knows the firm needs this program to build morale, particularly at the small stores in outlying locations. She also knows that a comprehensive program will bring her firm the greatest profit.
- *Paragraph #3*—This list of the five award categories is basically "for the record." It prevents later disagreement about what groupings were discussed. With agreement on these categories, Orley may assign her design staff to do some preliminary work on drawings and prices.
- *Paragraph #4*—In the meeting, Seuss expressed concern about the quality of the logo reproduction on the awards, because he knows that the company's founder and president (who designed the logo in 1939) will tolerate no departures from original specifications. The letter reassures Seuss that he can approve the reproduction and the model before any money is paid.
- *Paragraph #5*—Because the RFP will be forthcoming and Orley will surely be on the list, there's no need for her to call the customer. Indeed, this is a good time for her to back off and hope that all her efforts will pay dividends after the proposal is in. Orley ends by suggesting that Seuss give her a call if he

Uses the modified block style with a five-space paragraph indention

Connects with the previous meeting and makes the letter's purpose clear

Focuses on the customer's main concerns: employee morale and loyalty

Specifies, for the record, the tentative awards already agreed upon

Helps the customer plan the RFP

Gives assurance that the logo quality will be up to standard

Avoids offering to call, because in this case it might appear pushy or inappropriate

Signs first name only, because she has established a close tie with the reader

BEST AWARDS, INC.
707 First Avenue, Bee, Maine 04285
(303) 311-3533

June 4, 2000

Mr. Jay Seuss, Manager of Human Resources
Caroline Department Stores, Inc.
25 King Street
Charte, Vermont 05827

Dear Jay:

The three of us made real headway last Monday in defining the kind of service award program you want. I'd like to summarize our work so far and answer a question you posed about the logo.

You and Hal obviously feel strongly that improved service awards can help build morale among Caroline's 3,000 employees throughout the Southeast. Awards can help build and maintain loyalty among employees who work far from the corporate headquarters.

Your preferences for employee service awards, as I understand them, can be placed in five general service groupings:

* 5 years—gold pen and pencil (with logo)
* 10 years—ruby logo pin and $1000
* 15 years—emerald logo pin and $1500
* 20 years—diamond logo pin and $2000
* 25 years and over (5-year increments)—special gifts

The design and price range of the pins and the special gifts would need to be addressed in any proposals you solicit from my firm or others.

Jay, as for your concern about reproducing the logo, our designers always work from the precise specifications you have on file. In fact, we ask you to sign off on the quality of the reproduced logo before we proceed to develop a master and, of course, before you make any payment. That way you're guaranteed of receiving the logo quality you expect.

From what you said in our meeting, you will soon select the gift groupings and request bid proposals from Best and perhaps other firms. In the meantime, please call me if you need any more help preparing the request for proposal.

Sincerely,

Georgia

Georgia W. Orley
V.P., Northeast Region

vh

FIGURE 6–4
Letter after a meeting—modified block style, paragraph indention

needs any more help developing the RFP. It can be weak strategy to make a "call me if you have questions" type offer because you shift control to the reader. However, it's appropriate here, because writer control has been established in other ways.

Example #4: Letter of Reacquaintance (Refresher Letter)

This letter, shown in Figure 6–5, is written in the simplified style with the following features:

- No salutation
- A subject line in full capitals
- The reader's name included in the first line or two
- No closing
- Indenting much like that of the block style

This style has a less formal appearance, and it's useful when the reader's name isn't known. However, this style can appear too informal and impersonal. When you're not certain about your reader's stylistic preference, use the block or modified block styles.

Your best opportunity for business often flows from the good work you've already done. Statler is wise to write Swartz while the alarm project is still fresh in the customer's mind, especially with that association being a good one. Statler hopes that Swartz will be receptive to what's being offered in this letter because of the good will left over from that prior work.

The following is a critique of this letter:

- *Paragraph #1*—The writer wastes no time reminding the reader of the successful work performed last year but doesn't belabor the point. Instead, he moves quickly to the purpose of the letter—to describe Allied's new asbestos service. Using the word *asbestos* to someone charged with the safety of many people will certainly capture interest. Statler does so without resorting to needless fear mongering.
- *Paragraph #2*—Here Statler relates the problem to the Jessup County School System. First, he notes that many of the schools were built when asbestos was used in construction. Then he explains that the material can affect those within the structures long after construction.
- *Paragraph #3*—This paragraph's first sentence makes clear what Allied can do for Jessup: a diagnostic survey and, if necessary, cleanup. It's important that Statler convey the sense that Allied can solve the whole problem.
- *Paragraph #4*—Statler reinforces an already established theme of the professionalism that Allied brings to its jobs. In fact, this theme is the central selling point of the letter. He also uses the phrase "peace of mind" to convey the need for immediate action to resolve the potential problem.
- *Paragraph #5*—The last paragraph uses two good techniques. First, it mentions that an enclosed brochure contains further information. Second, it refers to an upcoming phone call as something that will serve the client. He wants to help Swartz identify and solve the problem, not just get a contract from the school system.

ALLIED SAFETY ENGINEERING
388 Kister Avenue
Seebrook, Oregon 97028
(808) 344-5646

August 21, 2000

Uses the simplified style with no salutation and closing

Mr. James Swartz, Safety Director
Jessup County School System
1111 Clay Street
Smiley, Oregon 97666

Uses a subject line to gain attention

NEW ASBESTOS ABATEMENT SERVICE NOW AVAILABLE

Refers to previous successful work; leads in naturally to the letter's subject (asbestos abatement)

We enjoyed working with you last year, James, to update your entire fire alarm system. Given the current concern in the country about another safety issue, asbestos, we wanted you to know that our staff now does abatement work.

Avoids alarmist tone; stays objective and informative

As you know, many of the state's school systems were constructed during years when asbestos was used as a primary insulator. Now we know that asbestos can cause illness and even premature death for those who work in buildings where asbestos was used in construction. Even a small portion of asbestos produces a major health hazard.

Comforts by showing how problems can be discovered and solved

Fortunately, there's a way to tell if you have a problem: the asbestos survey. This procedure, done by our certified asbestos-abatement professionals, results in a report that tells if the buildings are affected. And, if we find asbestos, we can remove it for you.

Reinforces the relationship between the writer's and reader's organizations

Jessup showed real foresight in modernizing its alarm system last year, James. Your desire for a thorough job on that project was matched, as you know, by the approach we take to our business. Now we'd like to help give you the peace of mind that will come from knowing that either (1) there is no asbestos problem in your 35 structures or (2) you have removed the hazardous material.

Makes brief reference to the enclosure; stays in control by mentioning he'll make a follow-up call

The enclosed brochure outlines our asbestos services. I'll give you a call in a few days to see if Allied can help you out.

Ben

Benjamin R. Statler
Project Manager

nb
Enclosure

FIGURE 6–5
Letter of reacquaintance (refresher letter)—simplified style

Example #5: Transmittal Letter (Cover Letter)

This letter, shown in Figure 6–6, is written with a block style, although any of the previously used styles would have been acceptable.

The transmittal letter introduces and supports the proposal it accompanies. It can be sent separate from the proposal or included as part of the proposal document. In this example, it was sent detached from the proposal. (The example is how a transmittal letter might have been written for the formal proposal model in Chapter 7.)

The following is a critique of the letter:

- *Enclosure Line*—The letter clearly identifies the name of the proposal—using the same title on the proposal cover and title page—and the customer RFP that generated the proposal.
- *Paragraph #1*—The focus is on the hull cleaning and maintenance needs of the customer and identifies the solution—the Seasled system. It also shows that the solution isn't based only on the RFP requirements but also on frequent dialogue between Hydrotech and the customer. Notice that the letter doesn't begin with banal comments such as "we are pleased to submit . . ." or "we welcome the opportunity to submit . . ." The customer can probably guess that you would be "pleased" or would "welcome the opportunity" to submit a proposal, especially if it results in a contract.
- *Paragraph #2*—Details about the Seasled solution are presented, including key features and benefits that will help the customer meet its needs. If just the features had been described, it would have been up to the reader to deduce the benefits.
- *Paragraphs #3 and 4*—Claims of product benefits can be unpersuasive if the reader is given no proof to back up those claims. This part summarizes the outstanding results customers have reported after using the Seasled service and then links these results to the ability of the Seasled service to meet the needs of Standard Shipping International.
- *Paragraph #5*—Although this is a short paragraph, it closes with three goals: (1) maintain control over the next step, (2) identify the point of contact at Hydrotech, and (3) stress the company's interest in gaining the contract. The advance notice to discuss the demonstration gives Wilson an important reason to contact the customer. A supervisor could have written and signed the letter; however, it was thought that if the president did it, the letter would reflect a stronger management commitment and interest in the new business. But, even with his signature and involvement, Wilson, the president, talked with the proposal manager to learn the main selling points in the proposal and then wrote the letter to support them.

Uses a block style

Uses an enclosure line to show the name of the proposal and the RFP to which it responds

Shows an understanding of customer needs and indicates the source of this knowledge

Summarizes the major features and benefits of the proposed hull cleaning service

Substantiates the cost-saving benefits of the service with the results experienced by customers

Maintains control of the next move and identifies the Hydrotech point-of-contact

With a letter from the company's president, shows corporate interest and commitment

Hydrotech Diving and Salvage, Inc.
Industrial Complex
New Orleans, Louisiana 70146

February 24, 2000

Ms. Susan Bard Jackson
Standard Shipping International
Fleet Drive
New York, New York 10019

Enclosure: "A Proposal for Hull Cleaning and Inspection" in response to RFP #SSI 02-07999, of February 1, 2000.

Dear Ms. Jackson:

Improved hull cleaning and maintenance for your galaxy-class oil tankers are operational and financial necessities for your company. As you stressed in the RFP and our frequent conversations, traditional approaches to these activities have become costly and inefficient. This proposal offers Hydrotech's innovative and affordable solution to your needs—the Seasled system.

Our proprietary Seasled system is on a diver-operated, self-propelled water vehicle for hull cleaning and maintenance. It is a safe, reliable, and effective system that can be quickly deployed to a variety of world-wide ports and open sea anchorages. Its use will reduce your hull cleaning costs and enhance the operational efficiency of your fleet.

Since the system became operational two years ago, we have cleaned more than 100 tanker hulls for our customers. The Seasled system has shown that it cleans and paints hulls about 50 percent faster than manual diver approaches. Our system has reduced hull cleaning expenses of our customers by 40 percent and has increased the annual at-sea availability of their ships by 10 percent.

Increased availability of your fleet will keep your ships where they can earn money—at sea—serving your customers, not tied up undergoing cleaning and repair. Hull cleaning retards the accumulation of marine growth on your hulls, allowing your ships to cruise faster with less fuel consumption.

I will call you next Wednesday to discuss our proposal's offer to conduct a personal demonstration of the Seasled system. Meanwhile, please contact me directly at 1-800-455-0590 to discuss any aspect of the proposal. We want your business and are committed to providing you a more efficient and less costly approach for your hull cleaning and maintenance program.

Sincerely,

William Bailey, Jr.

President and CEO, Hydrotech Diving and Salvage, Inc.

sf

FIGURE 6–6
Transmittal letter (cover letter)—block style

RULES FOR DEVELOPING EXECUTIVE SUMMARIES

> *Every* proposal you write needs an executive summary or proposal overview. A proposal of only a few pages can include this summary in the cover letter, but in all other cases you need a separate section. The executive summary is more than window dressing. For at least some of your readers—very likely including the final decision-makers—the executive summary will be the *only* part of your proposal they read.
>
> (Svoboda and Godfrey, 1989, p. 109)

Develop the executive summary to summarize your proposal and to highlight its key selling points. Unless otherwise directed by the customer, always prepare an executive summary because of its potential value to your reader *and* your proposal team.

Like a sales letter, it provides abbreviated support for the sales proposal. The executive summary can be submitted as one volume in a multivolume proposal or as a section in a single-volume proposal. For a multivolume proposal with a dedicated volume for the cost offer, also consider including the executive summary as part of the cost volume. This can allow the cost evaluators to read the executive summary if they don't have access to the noncost volumes and provide them a background understanding of the factors that drove the proposed cost. (Expect the U.S. government to restrict what its cost evaluators can read during a proposal evaluation.)

The executive summary can allow you to write more creatively and informally than you do in the main proposal, because readers are more likely to expect the summary to have an obvious sales pitch, presented in an entertaining and visually appealing layout (referred to in the proposal business as "pizzazz"). Use the following rules to develop your executive summary.

■ *Summary Rule #1: Start the process early.*

Assign a writer to plan and develop drafts of the executive summary as you would for other proposal sections. Start it early in the proposal process, although some will argue that the main proposal must be written before the executive summary can be written. The following are advantages in starting the executive summary early:

- Early assignment gives its writer time to develop and understand the key points of the main proposal and to plan, write, review, and revise it as you would do for other proposal parts. Although early drafts of the executive summary may be incomplete because details are missing in the main proposal, those details can be added to the executive summary as the main proposal matures. If you wait until the last minute to plan and write the executive summary, you may not have enough time to improve it with reviews and revisions.

- Its development can guide and reflect the development of details in the main proposal. Because an executive summary needs details, if it's vague, incomplete, or unsubstantiated, it may mirror a main proposal with the same problems. Thus, the executive summary can be a catalyst for resolving these problems.
- Its drafts can provide other proposal authors an overall view of the proposal themes and baselines, thereby promoting continuity and consistency among the details in the main proposal.

Summary Rule #2: Select a development plan and follow it.

To avoid the frustration of frequent restarts and rewrites, select a structure and content approach and then implement it. The following are four ways to develop your executive summary:

- Abstract—A condensed version of the main proposal, it addresses the major points of the main proposal in the order in which they appear. It reads as an abstract or miniversion of the main proposal and prepares the reader for the organization and content of the main proposal.
- Selection/evaluation criteria—Organized by the proposal grading criteria in the RFP, it uses the criteria topics as subheadings and summarizes your proposed approach to meeting the criteria requirements.
- Proposal themes—Organized by the major proposal themes (selling points) of your main proposal, it amplifies each theme with the benefit(s) it provides and backs up any benefit claims with credible proof.
- Customer instruction—It is organized and developed according to instruction from the customer. This guidance can be provided orally or by a written RFP.

Regardless of your development plan, avoid putting information in the executive summary that isn't addressed in the main proposal. If it's important enough to be in the executive summary, it's important enough to be in the main proposal. Following this guidance can be crucial in proposals to the U.S. government. An RFP from the federal government may require the input of an executive summary that won't be formally graded with the main proposal during the source selection process. In this case, if you put information only in the executive summary, its absence in the main proposal may mean the information—regardless of its value—will have no effect on the formal grading of your proposal.

> The best talent should be applied to preparing the executive summary. It should be written by a technical, management, or marketing person who has an overall view of the proposal. It should not be a team effort, but should be written by one person so that it has a consistent flow and style.
>
> (Hill, 1987, p. 167)

■ *Summary Rule #3: Prepare it as a sales document for various readers.*

Despite its name, the executive summary isn't necessarily for the executive-level reader who wants to review quickly the major points of the proposal. It can also be used by specialists or lower-level managers who would prefer more details in it than just a summary description of your product or service. For some, it might be the only part of the proposal they read, but, for others, it might be used to supplement their reading of the main proposal. Therefore, write the summary for any reader in your audience who might be involved in evaluating your proposal, and, as space permits, provide different levels of detail to meet the needs of various readers.

■ *Summary Rule #4: Make an early decision for its page allowance.*

There's no firm rule about how long an executive summary should be. Unless otherwise directed by the customer, follow a general rule of allocating pages equal to 5 to 10 percent of the total pages of the technical and management portions of the main proposal. For a short formal sales proposal of 25 pages, a one-page executive summary might be appropriate. However, a 35-page executive summary might be warranted for a lengthy, multivolume proposal.

Its length can be affected by many factors:

- Chosen format—As previously mentioned, there are different ways to organize the executive summary. Your chosen approach can greatly affect how many pages you'll need. For example, the abstract executive summary for a multivolume proposal will probably require more pages than a proposal theme summary that just needs to focus on several major proposal themes for the same proposal. Or, for a selection/evaluation criteria approach, you can expect the executive summary to have more pages for an RFP with 20 evaluation criteria than it would be for an RFP with six.
- Customer limitation—The RFP may specifically limit its page size or dictate that its pages be counted as part of an overall page limitation for the entire proposal or a specific proposal volume. Comply with the RFP instructions. If the length of the executive summary needs to be coordinated with the page allocation for other parts of the proposal, you must plan carefully. For example, if the main proposal, including the executive summary, is limited to 100 pages, you obviously shouldn't use 45 pages for the executive summary, because this allocation will not allow enough space for your main proposal to adequately address its applicable RFP requirements.
- Comparative value—The length of your executive summary can also be based on its comparative value to other parts of the proposal. For example, let's consider an RFP that requires an executive summary and includes its length as part of page allowance for the entire proposal—but doesn't include it as a factor in

the grading of your proposal. You then must decide how many pages of your allocation should be given to an executive summary, which will have no formal impact on the scoring of your proposal.

With its page allocation set, you can then realistically make an outline and the storyboard(s) for your executive summary.

■ *Summary Rule #5: Use a high percentage of graphics.*

As a general rule, a longer executive summary should devote at least 50 percent of its content to graphics. Use storyboard planning to identify executive summary graphics that help you summarize key proposal points in an interesting and aesthetic presentation. If readers must wade through executive summary pages saturated with text, they may lose interest. Graphics can allow your readers to scan the executive summary for the major selling points, benefits, and conclusions you want them to remember. Compared with text, a graphic can more effectively convey your message and require less space to explain a complex topic.

SUMMARY

Sales letters are any correspondence you send to the customer during the proposal process, including before and after proposal submittal. These letters have three main purposes: (1) to reinforce points already mentioned on the phone or in person, (2) to follow up questions asked by the client, and (3) to exercise control during the proposal process.

Use the following rules to write sales letters:

1. Use the "easy-in, easy-out" rhetorical pattern.
2. Use the 3-Cs pattern to gain and keep reader interest.
3. Capture interest with a strong first paragraph.
4. Convince readers in the body paragraphs.
5. Control the next move with the last paragraph.
6. Break up body paragraphs for easy reading.
7. Add a postscript for special attention.
8. Write one-page letters, using attachments for details.
9. Use the letter style your readers prefer.
10. Edit carefully.

Model examples were shown for the following five types of sales letters:

1. Letter of first contact, also known as a cold call letter—with block style
2. Letter after a phone call—with modified block style, no paragraph indention
3. Letter after a meeting—with modified block style, paragraph indention
4. Letter of reacquaintance, also known as a refresher letter—with simplified style
5. Transmittal letter, also known as a cover letter—with block style

Unless otherwise directed by the customer, always submit an executive summary with your proposal. It can be a valuable tool in developing and understanding the main proposal it supports. Follow these guidelines when developing the executive summary:

1. Start the process early.
2. Select a development plan and follow it.
3. Prepare it as a sales document for various readers.
4. Make an early decision for its page allowance.
5. Use a high percentage of graphics.

EXERCISES

Comply with the following procedures to establish a context for performing Exercises 1–6 that follow.

- Select a product or service with which you are reasonably familiar, on the basis of your work experience, research, or interests.
- Assume that you are responsible for marketing this product or service to Mr. Harry Black of ABC, Inc.
- Assume that Black is aware of your firm but hasn't done business with it.
- Answer the audience-analysis questions in Chapter 4 regarding Black. Use real or fictitious information, depending on whether you have an actual person in mind.
- If you use real information, ensure that it's not proprietary information from any company.

1. Cold Call Letter

Write Black a cold call letter in which you mention a need, introduce your product or service, and suggest the next step in the proposal process.

2. Sales Letter after a Phone Call

In a phone call you made after the cold call letter, Black showed some initial interest in your product or service. In fact, you managed to set up a meeting for two weeks from today (he'll be out of town next week). Now write a follow-up letter in which you summarize the conversation, confirm the meeting, and offer additional information that'll keep his interest.

3. Sales Letter after a Successful Meeting

You met with Black, and the meeting was a success. He wants you to give him a proposal within the next month. You plan to submit it in about three weeks. Now write a follow-up letter that (1) thanks him for the chance to submit the proposal, (2) mentions the planned submission date, (3) gives him more information about your ability to satisfy his needs, and (4) generally reinforces the good relationship that has been started.

4. Transmittal Letter for the Proposal

You have prepared a proposal for Black. Now write a transmittal letter to accompany and support the proposal. Among other points in it, stress the three main selling points of your product or service.

5. Sales Letter after Proposal Acceptance

Congratulations! After several meetings and one negotiation session, Black's firm has decided to award your firm a contract based on your proposal, and you will begin work soon. But writing a letter right now would even further convince Black that he made the right decision. Write this letter of reinforcement.

6. Refresher Letter Seeking Repeat Business

It has been six months since you successfully completed the job for Black and his firm, ABC, Inc. You know your best chance for business is with satisfied customers, so you decide to write Black again. Now write a refresher letter to offer him (1) the same product or service or (2) something new that you think he may need. Make sure the letter ends with a clear next step.

7. Executive Summary

In Chapter 7, there is an example of a formal sales proposal. Rewrite its executive summary with an abstract approach. Use whatever margins and line spacing you prefer, limiting its length to two pages.

7 Formal Proposals

OBJECTIVES

- **Learn the content of a formal proposal**
- **Learn how customers influence the content of formal proposals**
- **Learn rules to prepare for, influence, and use a request for proposal (RFP)**
- **Learn rules for developing formal proposals**
- **Examine a model of a formal sales proposal**
- **Perform exercises using chapter guidelines**

Up to this point, you've learned about the sales proposal process, bid/no bid decision, and strategy planning (Chapter 1); proposal management (Chapter 2); proposal content planning (Chapter 3); proposal writing (Chapter 4); proposal illustration (Chapter 5); and writing sales letters and executive summaries (Chapter 6). This chapter describes how this knowledge plus some additional techniques will help you to produce a responsive and winning formal proposal.

In general, proposals can be classified by the following definitions:

- Solicited (requested) or unsolicited (not requested) by the customer
- Directed to someone in your company (in-house) or in another firm (external)
- Submitted to sell a product or service (sales proposal), obtain funding for a research project (grant proposal), or secure support for an internal procedural change (planning proposal)
- Written in a short letter or memorandum form (informal) or in a longer and bound form (formal)

This chapter focuses on solicited and unsolicited formal proposals written for external readers.

FORMAL PROPOSAL FEATURES

Formal proposals contain more than five pages, excluding attachments. Typically, they are bound and contain many sections. A formal proposal is appropriate when one or more of the following conditions exist:

- The main body of the proposal requires more than five pages.
- The scope of the proposed product or service or the dollar amount of the potential sale justifies a formal response, showing the customer the importance you place on the potential contract.
- The customer prefers the formal proposal format.

After applying these criteria, if you're still in doubt about which format to use, write a formal proposal. It's better to err on the side of formality. Plus, the distinct sections of a formal proposal can make it more readable, organized, and visually impressive than an informal proposal.

The three major design features of a formal proposal—content, organization, and format—can vary greatly, depending on the targeted customer and whether the proposal is solicited or unsolicited:

- For a solicited proposal in response to a customer's request for proposal (RFP), you can expect the RFP document to provide at least general guidance about these features. The nature and extent of these guidelines can differ depending on the needs of the customer and whether the RFP is issued by a commercial or government source.
- For an unsolicited proposal, you aren't responding to an RFP and therefore don't have its instructions dictating how to outline, write, and format your proposal. Therefore, expect to have greater freedom in choosing the proposal design features. Even with this leeway, ask your customers (commercial or government) for guidelines they may have for unsolicited proposals. Also consider showing customers your proposal outline for their review and comment.

Whether a proposal is solicited or unsolicited, formal proposals often provide the following details:

- Purpose—This part describes the need or problem that the proposal addresses. Expect this detail to be in a section within a proposal volume.
- Technical approach—This part explains the technical features and benefits of the proposed product or service, including such details as design, reliability, testing, operation, logistical and training support, and quality assurance. Expect this detail to be in a section within a proposal volume or addressed as the topic of a stand-alone volume.
- Management approach—This part explains how the work will be managed to deliver the proposed product or service. Expect this detail to be in a section within a proposal volume or addressed as the topic of a stand-alone volume.
- Related experience/past performance—This part describes your past or current contracts and how this work demonstrates your ability to deliver the proposed

product or service. Expect this detail to be in a section within a proposal volume or addressed as the topic of a stand-alone volume.

- Cost information—This part shows the bottom line: how much will the product or service cost? It can also describe the cost of various options and how different factors can affect the final price. Expect this detail to be in a section within a proposal volume or addressed as the topic of a stand-alone volume.

SOLICITATION INSTRUCTIONS

Depending on what will be proposed and your relationship with the customer, the request to write a formal proposal may come in many forms:

- Phone call
- Comment during a meeting
- Short letter
- Written RFP

Regardless of the source, you need to learn what product or service is needed, the information to be included in the proposal, and how to organize and format the proposal. To show how written requirements can affect the development of a solicited formal proposal, let's look at some guidance you may receive from government and nongovernment customers.

U.S. Government Customers

Selling to the U.S. government is controlled by many regulations and procedures. The basic source for federal contracting guidelines is the Federal Acquisition Regulation (FAR), which is supplemented by procedures set by individual federal departments and agencies. The FAR is frequently revised and is easily accessible via U.S. government sources on the Internet. Although learning and complying with the FAR and its supplements can be challenging, they do provide a structured set of procurement standards. With the universal application of these guidelines, the U.S. government promotes a "level playing field" for those competing for federal contracts—unlike commercial customers, which have much more latitude in how they solicit and award contracts.

It's beyond the scope of this book to explain all federal procurement regulations and procedures. However, we'll describe how to find federal contract opportunities and then briefly address two ways to gain federal contracts: responding to a federal government RFP and completing a Standard Form 255 (SF255). RFP and SF255 responses are considered solicited. (For the most current guidelines about selling to the federal government, talk with your U.S government customers and refer to the latest FAR.)

Contract Opportunities. The *Commerce Business Daily (CBD),* a U.S. government publication, is a key source for monitoring large contract opportunities with the federal government. It announces pending procurements (supplies and services), awarded contracts that could lead to subcontracts, the sale of surplus materials, and opportunities for research and development contracts. Released daily, except on weekends and government holidays, it identifies contract possibilities and awards worth more than $25,000.

The *CBD* provides the following information:

- Pending release of draft and formal RFPs and related plans for preproposal bidder conferences—These advance notices describe how to obtain copies of the documents and can help you plan your proposal marketing and development effort. To receive the solicitation, comply with *CBD* instructions. RFPs can be obtained in print through the U.S. mail or in an electronic format from a government web site on the Internet.
- Awarded contracts and the winning contractors—This information can indicate the success (and failure) of others in your industry. It can also identify companies that might use your firm as a subcontractor to perform the awarded contract. If your small business seeks subcontracts for federal contract or subcontract work, register on the Procurement Marketing and Access Network (PRO-Net), a database compiled by the Small Business Administration (SBA). The PRO-Net is accessed by prime commercial contractors and the U.S. government to find small businesses suitable for federal work.

The *CBD* is available in printed or electronic form. The U.S. Government Printing Office provides subscriptions for the printed version through the U.S. mail. It is also available in print at local public libraries and electronically from commercial and U.S. government sources on the Internet.

The electronic version of the *CBD* provides two key benefits:

- It permits access to *CBD* information on the afternoon before the publication date, sooner than you can receive the printed information by mail.
- It allows you to perform automatic word searches to find specific contract opportunities.

Commercial services will monitor the *CBD* for your areas of interest and alert you when they appear. If you seek contracts from state and local governments, determine how they announce contract opportunities through printed or electronic (Internet) media and take advantage of these sources.

Despite the importance of the *CBD* for large procurements, a significant amount of federal government work is awarded for products and services worth less than $25,000. Therefore, details about these opportunities don't appear in the *CBD.* Most federal procurements under $25,000 are awarded based on simplified or sealed bidding procedures—not a formal proposal response to an RFP. To learn about these opportunities, maintain close contact with your target government customers. Read their bid boards that announce pending contract needs and ensure that you're placed on their bidders' list, so you can be contacted about these opportunities. To be placed on a bidders' list, you may be required to complete and submit a Standard Form 129 (SF129), "Solicitation Mailing List Application," to the government agency. You can also use the SF129 to get placed on mailing lists to receive RFPs for procurements worth more than $25,000.

Federal Government RFPs. These solicitation documents contain the proposal requirements for winning major competitive and negotiated contracts from the U.S. government. The resulting contract value is usually more than $25,000. Some agencies are raising this ceiling to $100,000 due to recent revisions to the FAR. You can expect to have at least 30 days after RFP release to submit the proposal. The RFP will dictate the delivery deadline (date and time) and location. Although RFPs can vary greatly in size and level of detail, they contain the sections shown in Table 7–1.

For the proposal writer, the most important RFP sections are A, L, M, and C. Figure 7–1 shows how these sections influence the development of your proposal. The following describes these RFP parts in more detail:

- Section A—This section indicates when and where to deliver the proposal, who you should contact regarding the RFP, and what the RFP contains. Ensure that your copy of the RFP has all the parts shown in this section's table of contents.
- Section L—This section dictates the basic organization and content of your proposal. It also states requirements for the length, format, and typography of the proposal. Follow these directions to the letter. In other words, don't disobey its instruction by changing the section numbers and titles it gives for proposal headings, don't substitute the required proposal outline with one you think "flows" better, and don't submit more pages than allowed.
- Section M—This section states criteria the government will use to evaluate all submitted proposals. The amount of detail divulged about the criteria and their relative weighting on the evaluation grade will vary among RFPs. This section identifies the key customer concerns the proposal needs to address. Look for a direct correlation between the topics requested in Section L and the evaluation criteria in Section M. Section L provides the basic framework and content of the proposal sections, while Section M amplifies or supplements the content. Therefore, when you outline the proposal based on Section L, ensure that the evaluation criteria are reflected within the body of the outline.
- Section C—The Statement of Work (SOW) in this section provides the technical and management details of how the customer wants the proposed product or service to be designed, produced, delivered, implemented, or operated. Section C requirements need to be addressed within the outline developed from Sections L and M. The SOW can have attachments for providing more detailed requirements.

The information above doesn't mean the proposal writer can ignore the other RFP sections. For example, Sections B, F, H, and I can also directly affect the content of your technical and management proposal sections—although your finance and contracts staffs may have more interest in these sections than the team writers.

> The Government is often in the same position as an ordinary shopper, needing something but not knowing what's available. When an RFP is issued, the Government is surveying the market. Just like someone shopping, it may know that it needs a stereo but doesn't know how much a stereo costs or what features can be obtained.
>
> (McVay, 1987, pp. 59–60)

TABLE 7–1: Sections of a Federal Government RFP

RFP Section	Description	Key Information
A	Standard Form 33 (SF33) solicitation/contract form	RFP table of contents, customer contact information, due date of proposal, and where the proposal is to be submitted; note: requires some information and a signature from the offeror; submitted with proposal; serves as a contract when signed by offeror and government
B	Supplies or services and prices/costs	Contract Line Item Number (CLIN) list to specify what product/service items are required and how they are to be priced
C	Description/specifications/work statement	Definition of required work and products/services; presented in a Statement of Work (SOW) or a referenced attachment (e.g., a technical specification)
D	Packaging and marking	Product packaging and marking requirements
E	Inspection and acceptance	Inspection and acceptance criteria for the product/service
F	Deliveries or performance	Delivery dates for the product/service
G	Contract administration data	Miscellaneous contract administration requirements
H	Special contract requirements	Contract requirements not included in Section I
I	Contract clauses	Contract requirements in the form of standard contract clauses from government regulations; provided in words or by reference (number and title) to the clause
J	List of attachments	Exhibits and attachments included for the applicable RFP; can include the Contract Data Requirements List (CDRL) for all data to be delivered to the government, Data Item Descriptions (DIDs) that explain the content and format of the required data in the CDRL items, and technical specifications
K	Representations, certifications, and other statements of offerors	Representations and certifications to be accepted, acknowledged, and signed by the offeror; signed sheets of acknowledgment or acceptance to be submitted with the proposal
L	Instructions, conditions, and notices to offerors	Instructions for proposal preparation, including proposal format, organization (outline), content, and size
M	Evaluation factors for award	Description of how the proposal will be evaluated

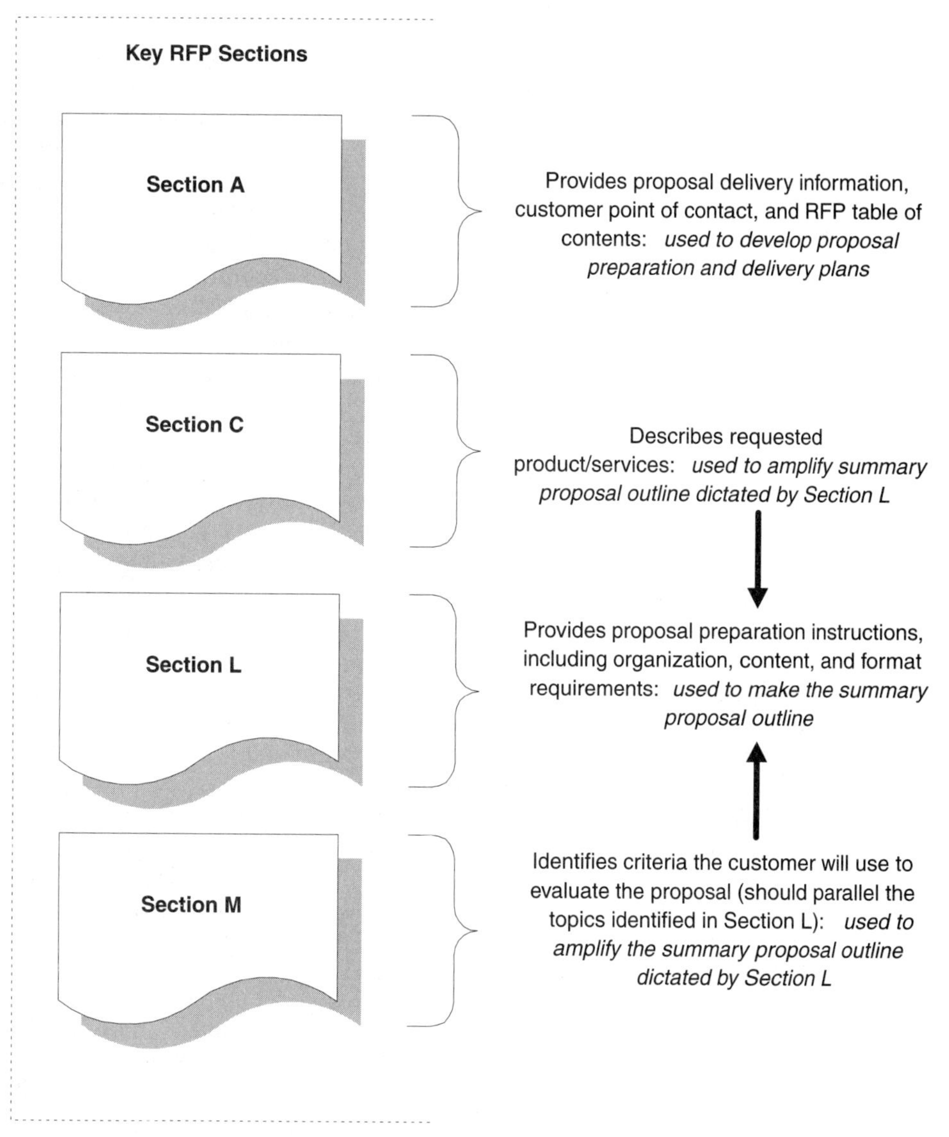

FIGURE 7–1
Key sections of a U.S. Government RFP

Standard Form 255 (SF255). This form, "The Architect-Engineer and Related Services Questionnaire for Specific Project," is used by various federal and local government organizations to procure architectural/engineering (A/E) services. The *CBD* announces federal A/E opportunities and describes the services and information that must be provided in the SF255 submittal. This announcement is made once for an A/E opportunity. After it appears, expect to have 30 days to submit the completed SF255.

The SF255 isn't submitted with a cost proposal. Therefore, the *CBD* synopsis and the SF255 aren't RFPs, and the SF255 submittal isn't a formal proposal. The cost of the offered services is proposed and negotiated later, after the U.S. government has trimmed the list of qualified firms (called "short listing") based on its evaluation of the submitted SF255 forms. After the short listing, the government interviews the remaining competitors and chooses the best candidate. The top-ranked contractor then submits the required costing information to the government for negotiation. If this contractor and the government cannot agree on contractual terms—including the price of the services—the government begins negotiation with the next ranked candidate.

Although the SF255 submittal isn't technically a formal proposal, the form's content is similar in scope to that of a formal proposal. Its planning and development are also common tasks of a firm's proposal organization. For these reasons, the SF255 is described in this chapter. Table 7–2 lists the information required in the SF255, which challenges a company to show that its experience and staff are indicators that it can perform the requested services.

The SF255 is a supplement to the Standard Form 254 (SF254), "Architect-Engineer and Related Services Questionnaire." Whereas the SF255 data is centered on a particular A/E need, the SF254 provides a general summary of your experience and resources. For example, the SF254 doesn't include the names or résumés of key personnel who will be assigned to a specific project. The SF254 is sent by firms to agencies that are potential sources of A/E service contracts. It can be used by the agencies to support their review of SF255 submittals or to identify candidate firms for A/E contracts without the use of SF255 submittals in the selection process. The SF254 should be updated annually or whenever there are significant changes. A current SF254 should accompany any SF255 submittal.

> Deciding to pursue a project with the Department of Veterans Affairs, an A/E firm developed its submittal strategy. Previous experience with the agency led the team leader to downplay the importance of interior design services and staff even though they were specifically mentioned in the *CBD* announcement. "They always mention interiors and we always include them, but it's always such a small part of the project. These other elements are more important to this project." Every project is different. By presuming to understand the agency's intent and second-guessing the announcement, the firm excluded itself from contention.
>
> (Usrey, 1996, p. 50)

Although both forms are available in print from the U.S. government, A/E firms commonly use electronic (computer file) versions of the form. The forms can be easily made in template form with popular word processing programs or can be purchased in commercial software packages.

Grant Providers

Grant proposals are written by for-profit and nonprofit organizations. A common goal of the for-profit grant proposal is to gain funding for research and development—especially from the U.S. government. The nonprofit grant proposal often seeks

TABLE 7–2: SF255 Item Description

SF255 Item	Requested Information
1.	Project name/location for which the firm is filing
2.	*CBD* announcement date (if any)
2a.	Agency identification number (if any)
3.	Firm (or joint-venture) name and address
3a.	Name, title, and telephone number of principal to contact
3b.	Address of office to perform work, if different from Item 3
4.	Personnel by discipline
5.	If joint venture, a list of participating firms and an outline of responsibility (including administrative, technical, and financial) for each firm
5a.	Indication (yes or no) of whether the joint venture in Item 5 has previously worked together
6.	If not a joint venture, a list of outside key consultants/associates anticipated for the project: • Name and address • Specialty • Whether the consultant/associate has previously worked with the prime
7.	Brief resume of key persons, specialists, and individual consultants anticipated for the project: • Name and title • Project assignment • Name of firm with which associated • Years experience (with firm, with other firms) • Education—degree(s)/year/specialization • Active registration—year first registered/discipline • Other experience and qualifications relevant to the proposed project
8.	Work by firms or joint-venture members which best illustrates current qualifications relevant to the project (no more than 10 projects): • Project name and location • Nature of firm's responsibility • Project owner's name and address and project manager's name and phone number • Completion date (actual or estimated) • Estimated cost in thousands (entire project and work for which firm was/is responsible)

continued

TABLE 7–2: —*continued*

SF255 Item	Requested Information
9.	All work by firms or joint-venture members currently being performed directly for federal agencies: • Project name and location • Nature of firm's responsibility • Agency (responsible office) name and address and project manager's name and phone number • Percent complete • Estimated cost in thousands (entire project and work for which firm is responsible)
10.	Any additional information or description of resources (including any computer design capabilities) supporting your firm's qualifications for the proposed project (Note: The broad nature of this item allows you to address many topics in this portion of the form. For example, it can be used to explain your project management and quality assurance approaches, highlight the major "selling" features of your A/E services, amplify the applicability of current or past contract work, and explain how you meet the evaluation criteria cited in the *CBD* announcement. The evaluation criteria are commonly used to outline the item's content. Your response to the item may be page-limited by the customer.)
11.	Verification that the form is a statement of facts: • Signature • Typed name and title • Date

funding or in-kind support (such as equipment, facility usage, or volunteer assistance) for projects concerned with social and medical needs in our society.

It's beyond the scope of this book to describe fully the source and content of all grant proposals. However, we'll describe some features of nonprofit grant proposals to further explain the development and content of formal proposals. (For more guidance about how nonprofit organizations should develop grant proposals, contact the Foundation Center, 79 Fifth Ave/16 Street, New York, NY 10003, phone 212-620-4230, or access its Internet web site. It's a nonprofit source of information about corporate and foundation philanthropy.)

The existence of many nonprofit organizations depends on their success in writing formal proposals to obtain grants from foundations (corporate and private) and governments (state and federal). Although their content, format, and organization can vary, nonprofit grant proposals can be expected to have (1) an executive summary, (2) a statement of the problem/need, (3) the approach for addressing the problem/need, (4) the project objectives, (5) the project management approach, (6) the project cost, and (7) the approach for determining if the project objectives are being met.

As for other formal proposals, it's important to learn how the targeted grant source (the customer) wants the proposal written and when it should be submitted for consideration. Effective grant proposals require the same research, planning, writing, and selling techniques other formal proposals need. However, the following tasks for grant proposal writing deserve the special attention of nonprofit organizations:

- Researching to find potential grant sources and to learn about their grant and proposal requirements
- Convincing these sources that your proposal addresses a problem/need (objective) that deserves their support
- Showing that you have a sound approach for meeting the objective and that you'll be able to measure the success of your approach

> Be sure to check the foundation's instructions for how and when to apply. Some foundations will accept proposals at any time. Others have specific deadlines. Foundations will also differ in the materials they want a grant application to submit. Some will list the specific information they want and the format you should adopt. Others will have an application form. . . . Whatever the foundation's guidelines, pay careful attention to them—and follow them.
>
> (Geever and McNeill, 1993, pp. 67–68)

Research. There are thousands of potential funding and asset sources for nonprofit organizations. The challenge is to find the foundation or government source that shares and is willing to support the goals of your nonprofit organization.

One of the best ways to find candidate grant sources is through the many publications of the Foundation Center. These publications can help you find private and corporate foundations that share the same interest (issue and geographic focus) of your nonprofit organization. They describe foundation procedures, requirements, and preferences for grant proposals. They also provide information about the foundation points of contact, officers, history, publications, financial status, and grant-making patterns. The publications can be obtained from the Foundation Center or various public libraries in the United States.

Problem/Need. When you have targeted a potential grant source, you must convince the foundation that the addressed problem/need deserves its support—particularly when the foundation provides only a general description of the support it's willing to give. Therefore, a nonprofit grant proposal must clearly define the problem/need and explain how it relates to the foundation's concerns. Focus the grant proposal on a manageable portion of the problem/need, avoiding those that are too expansive and complex for you to resolve or mitigate with your resources and available time. A well-defined problem/need will help the nonprofit organization propose more specific objectives and approaches.

Unlike the proposal that responds to a stated need or problem in an RFP, the success of a grant proposal depends on how well it explains and substantiates the problem/need and then appeals to the foundation for a call to action.

Approach and Measurement. As with any formal proposal, the grant proposal must present an effective approach to meeting the objectives. Assuming you have a sound approach to address the problem/need, the grant proposal must explain the objectives and how it will be determined that the objectives are being met.

A foundation wants to know that its resources will be used effectively and that your organization will be accountable for its work. To address these concerns, explain your procedures for monitoring project performance, including schedule and cost results, and state how these results will be reported to the foundation. At the core of this planning should be a realistic project schedule and budget, supported with milestones defined with specific objectives. Measuring your performance will allow you to resolve problems that could prevent you from meeting the grant objectives.

Other Customers

Commercial and nonfederal government customers can use many types of written solicitations to request proposals from bidders. RFPs can vary greatly in how much guidance they provide for proposal content, format, and organization. The guidance can be driven by a simple list of questions or by a more detailed RFP, similar to a U.S. government solicitation.

A question solicitation provides a list of questions to be answered in the proposal. Unless the customer otherwise directs, your proposal should list the answers in the order the questions are asked. This approach simplifies the development of a responsive proposal outline and makes it easier for customers to match your answers to their questions. You can make it even easier to understand your answers by repeating the question before your responses.

Regardless of its complexity, a non-U.S. government RFP usually asks for a bidder's response to a statement of needs and requirements and provides at least some guidance for outlining the proposal. Let's consider what you might do to reply to the needs-and-requirements RFP shown in Figure 7–2. (This is a fictitious RFP.) After receiving this RFP, you proceed with the following planning steps:

- Decide to submit the proposal based on the bid/no bid factors, as described in Chapter 1.
- Assign proposal preparation tasks among the members of your proposal group as described in Chapter 2.
- Assemble and submit a list of questions to ask the customer to clarify any issues about the RFP, as described later in this chapter.
- Develop a proposal outline responsive to RFP requirements, as described in Chapter 3.

CITY OF CARBONDALE
P.O. BOX 1234
CARBONDALE, CALIFORNIA 99444

Request for Proposal
Service: Removing Asbestos from 10 City Buildings

1. Carbondale's Department of Public Works invites proposals from qualified consultants to (1) survey the degree to which asbestos is a health problem in 10 city buildings and then (2) coordinate the removal or containment of whatever asbestos is determined to be a hazard to public health. The purpose of the proposal is to select a consultant firm most qualified and interested in doing the work.

2. Without knowledge of the scope of the problem in the buildings, precise cost estimates are not expected. However, in the proposals, the responding firm should include a schedule of its usual charges for similar past projects. The city will negotiate a final fee with the most qualified firm, hoping that a mutually agreed-upon arrangement can be achieved. If such negotiations are not successful, the city will seek to negotiate with the second-choice firm. Proposals submitted for this project may be used to select professionals to complete other similar projects and may result in more than one contract with the city.

3. The entire asbestos program to be completed by the firm selected will involve three stages:
   - Stage 1: Evaluating the 10 sites
   - Stage 2: Writing a recommendation report with cost estimates
   - Stage 3: Completing the construction phase

4. Proposals should address the firm's ability to complete all three stages, making sure to indicate what subcontractors, if any, will be needed to complete Stage 3. Areas covered should include similar projects successfully completed, the specific people available in the firm who have done such work and their qualifications, and any certifications acquired by the firm or individuals. Of major importance are the safety procedures and techniques the firm plans to use in any necessary asbestos removal, with particular reference to concern for worker and occupant safety. Finally, some reference should be made to the number of people the firm could apply to the project, and the degree to which similar jobs have come in on schedule.

5. The two most qualified firms will be invited to deliver short presentations on the highlights of their proposals. Proposals are due in the office of Mr. James Schotsky, City Engineer, by 3:00 p.m. on October 20, 2000. Inquiries about this request should be made to Mr. Schotsky at 678-756-7676.

FIGURE 7–2
Needs-and-requirements RFP

You carefully analyze the RFP, so you can propose the best solution in the most responsive proposal package. You objectively assess your ability to satisfy your customer's needs, because to be competitive, your proposal should provide a responsive and best-value (not necessarily the lowest cost) solution to meet that need.

The RFP in Figure 7–2 contains five sections:

- Section 1—summarizes the purpose and scope of the requested service
- Section 2—summarizes the cost proposal and negotiation process for the contract award
- Section 3—briefly describes the three stages of the project that will provide the needed services
- Section 4—indicates what information is to be provided to describe the performance of the three stages
- Section 5—provides instructions for submitting questions and the proposal itself and explains follow-on activity after proposal submittal

You identify the customer's major concerns and needs and the effect they have on the proposal:

- Because one main customer concern is experience, you decide to summarize all your asbestos projects and emphasize work done for other metropolitan areas. This work will be described in terms that show a clear correlation between it and the proposed services.
- Another customer concern is personnel qualifications, such as degrees, certifications, short courses, and specific project experience. You decide to include updated and readable résumés for the assigned personnel, clearly showing how their background and qualifications are applicable to the work they'll perform for the proposed services.
- Another major concern is safety; it seems probable that the City of Carbondale is very aware of the dangers to workers and occupants from even minimal exposure to asbestos. The city wants assurance that the process will be rigorously monitored. You decide to provide an exhaustive list of procedures and then emphasize that your certified industrial hygienist will always be on-site. You also plan to stress your 100 percent safety record in your asbestos removal work and to describe your new isolation technique that allows the building to be occupied while the asbestos is being removed from it.

There are also some less pressing concerns; nevertheless, they are important to identify and resolve:

- Although the customer's main interest is the firm's qualifications, cost is also a concern. It appears from the RFP, though, that a firm won't be excluded because of cost, unless final negotiations fail. You decide to focus on the experience and skills of your personnel, and the effectiveness and safety of your operational services, to offset what you think may be a high cost of your services compared with that of your competitors.
- Although the RFP doesn't provide explicit instructions on an outline for your proposal content, it does suggest that the proposal should explain your ap-

proach to perform the three specific stages. You decide to devote a major proposal section to each stage, integrating the other requested information in those three sections or other supporting sections. Proposal details will focus on how you intend to meet the requirements of each stage.

- The customer wants to know what subcontractors, if any, will be used to help remove the asbestos. You decide to stress the reliability and experience of the subcontractors you have used on other jobs, as well as your plans for managing the subcontractors.

At this point, you probably have enough information to write an effective proposal; however, you decide that there are some questions you would like to ask Mr. Schotsky:

- Is there more information about the available budget for the solicited services?
- Is there a preferred format or style for the proposal?
- For what level of management and staffing should résumés be submitted?
- What kinds of buildings are involved and what are their ages? (Although the RFP seems to intentionally avoid such detail, this information could help you tailor your proposal for the proposed service.)

RULES FOR RFP SUPPORT AND USE

> The RFPs represent internal agreements that the client's staff makes, and then and only then, releases to contractors for bidding. The RFPs reflect political and financial battles won and lost, and their scopes of work are very dear to the people who will evaluate incoming contractor proposals. Therefore, the RFP dictates how the team should react, how the team should be structured, and what the writing plan should contain. . . . It is the rule book to which there is no appeal; any contractor who deviates from the RFP can summarily be dismissed from the competition just for failing to follow instructions.
>
> (Whalen, 1996, pp. 6–7 & 6–8)

Careful analysis of the RFP and other factors leads to a no-bid decision or to a commitment to develop the proposal. Use the following rules for reviewing, using, responding to, and possibly influencing the RFP.

■ *RFP Rule #1: Participate in draft RFP reviews.*

If a customer releases a draft RFP for industry review, take advantage of the chance to make comments, ask questions, or offer change recommendations. This process can benefit you and the customer for the following reasons:

- It can lead the customer to write a better RFP, including one better matched to your product or service.

- It can show your interest in helping the customer and your commitment to bidding for the contract.
- The review of the draft RFP can lead you to begin early proposal planning, including a customer needs assessment, a bid/no bid decision, and strategy planning. It can also help your team start a proposal draft based on a draft RFP.

Expect the opportunity to ask the customer questions about the RFP after it's formally released. The customer will typically give you a deadline to submit written questions. However, your best chance to influence RFP content is during its draft stage.

■ *RFP Rule #2: Submit comments, questions, and recommendations.*

Here are some recommendations for submitting your response to the customer about a draft or formal RFP:

- Formal process—Set a formal process for collecting, reviewing, screening, and editing responses before they go to the customer. Ensure that responses don't contradict or duplicate each other and that they are well written.
- Template—Use a template form to draft and review your responses internally before submittal. The form in Figure 7–3 can be used to review draft and formal RFPs. If the review input is approved for submittal, the form can also be used to edit the comment. Organize the forms with a log numbering system, referenced to the proposal section or volume. This logging system can help you and the customer to better track, reference, and process responses. After the form is approved for submittal, it can be sent in that format to the customer. But, if it is, be careful to delete the information used only by the proposal team.
- Purpose—Have a purpose for every comment or question. Don't nitpick typos and trivial errors or make comments just to show how smart you are. If you recommend a change to the draft or formal RFP, justify your idea by focusing on how it can help the customer. Look for ways to justify your recommendation with a win-win attitude—that is, it helps you as the bidder, as well as the customer.
- Need—Ask a question only when you really need the answer. First, by asking a question, you can expect that the answer will also be provided to your competition—help that you may not want to give. Second, if you fear what the answer could be, it might be best to write your proposal based on your best interpretation of the RFP and then hope you can later change your response if the customer takes exception to your interpretation. However, taking this approach about an ambiguous technical specification in the RFP might be a serious proposal mistake. Therefore, carefully analyze the ramifications of not asking an important question.
- Anonymity—Avoid naming your company in your written inputs, because customers often send their answers and all received responses to the potential bidders. If your competitors know your comments, questions, and recommendations—which can indicate your concerns, weaknesses, misunderstandings, and solutions—they could gain a competitive advantage. Obviously, don't put your company's

<table>
<tr><td>RFP REVIEW FORM</td><td colspan="3">Disposition/Comments</td></tr>
<tr><td>Master Log #:
(Master tracking #
for all inputs to customer—
Note #1)</td><td colspan="2">Accept:
(Indication that the response
will be sent to customer—
Note #2)</td><td>Reject:
(Indication that the response
won’t be sent to customer—
Note #2)</td></tr>
<tr><td colspan="2">Applicable RFP Section/Page:
(Specific referral to the RFP: paragraph,
section, and page numbers—Note #1)</td><td colspan="2">Topic:
(Title of the RFP section and/or summary
of the associated topic—Note #1)</td></tr>
<tr><td colspan="4">Comment/Question:
(A specific comment/question with any required background or amplifying information—Note #1)</td></tr>
<tr><td colspan="4">Recommendation/Rationale:
(A specific recommendation and justification for an RFP change or other customer action—
Note #1)</td></tr>
<tr><td colspan="2">Reviewer Name, Phone #, and E-mail Address:
(Note #2)</td><td>Date: (Note #2)</td><td>Reviewer Log #:
(Reviewer initials
and input number—
Note #2)</td></tr>
</table>

Note #1: The information is used for customer input.
Note #2: The information is used only by the proposal team.

FIGURE 7–3
RFP Review Form

name in the comment or question itself. Also delete anything else that could identify your company, such as a product description, company location, or contract history.

RFP Rule #3: Assign an RFP coordinator.

Assign a coordinator (a person or functional group) to monitor the pending release of RFPs and then to receive and appropriately distribute them. This coordination will reduce the chances of RFPs appearing unexpectedly from the customer, becoming lost in the external or internal distribution process, or being sent to the wrong people in your organization. Don't lose valuable proposal preparation time because the RFP gets sidetracked in the company mail or the wrong in-basket.

A central point for RFP receipt and distribution can also help businesses that have dispersed companies or branch offices within their corporate structure. A coordinator can ensure that the RFP is sent to the correct company or office and that different groups within the same corporation don't compete against each other by bidding on the same contract.

Ensure that your company meets all requirements to be placed on the customer's distribution list for RFPs. Regularly review sources that can alert you of pending RFP releases. For example, stay in frequent contact with current and prospective customers. Also, gather marketing information from such sources as the *CBD*, newspapers, trade magazines, and customer web sites.

RFP Rule #4: Control RFP configuration and distribution.

The RFP coordinator and proposal management must work together to ensure that the most current version of the RFP is distributed to the proposal team. Maintain a list of your RFP recipients, so you can quickly copy and distribute RFP amendments (customer changes or additions to the RFP) to those who need them.

RFP Rule #5: Participate in bidders' meetings.

Meetings between prospective bidders and the customer before and after formal RFP release are good ways to learn more about your customer and competition. These meetings are typically arranged by the customer.

The customer may ask prospective bidders for written questions to be submitted before the meeting and then provide oral and written answers for these questions at the meeting. Attendees may also be allowed to ask questions during the session. Written answers may be documented later as a formal RFP amendment and can initiate customer changes to formal RFP requirements.

Your company should participate in these activities. Ensure that the right people attend and give them specific objectives to achieve. Selecting the appropriate representatives can be especially crucial if the customer limits the number of attendees per company. Your representatives should have the following qualifications:

- An understanding of your objective in attending the meeting and the customer's purpose for having the meeting

- An understanding of questions, comments, and recommendations you submitted to the customer before the meeting
- An ability to identify and collect marketing intelligence, including customer needs not expressed in the draft or formal RFP and information about your competition
- Knowledge of the customer decision makers, the meeting leaders, and the customer requirements in the draft or formal RFP
- An understanding of your proposal strategy and the product or service you will propose

Avoid sending only marketing or sales representatives to bidders' meetings, when there may be more suitable attendees. For example, the best participants might be the proposal manager responsible for leading the proposal development or the project manager responsible for providing the product or service if the proposal is a winner. Attendees should provide the proposal team with an oral briefing and a written report, summarizing what was learned in the meeting.

At the meeting, your representatives should note which companies are represented. This could help you to identify your competition and to find companies willing to serve as team members or subcontractors in support of your proposal. When your potential competitors speak at the meeting, listen for anything that could be used for your competitive advantage. Caution your representatives to be careful about what they say at the meeting—especially about your proposal strategy or concerns you have about RFP requirements. (You can also identify other potential bidders by reading the distribution lists of correspondence, RFPs, and RFP amendments you receive from the customer.)

RFP Rule #6: Use requirements and compliance matrixes to be responsive.

What is meant by being responsive to the RFP? It means that your proposal offers a product or service that meets RFP requirements and provides information that meets RFP guidelines for proposal organization, format, and content. Don't lose a proposal competition because you inadvertently failed to meet the RFP requirements. Ensure that your proposal responds to all RFP requirements by using two matrixes:

- Requirements matrix—This matrix cross-references your proposal outline to applicable RFP sections. More fully described in Chapter 3, it's used for storyboarding and planning the content of your first proposal draft. Although primarily a planning document for writers, the matrix can also be used by review teams to verify that proposal drafts address all RFP requirements.
- Compliance matrix—This matrix, which can assist you and the customer, indicates where the RFP requirements are addressed in the proposal. It helps you verify that the proposal does indeed respond to all RFP requirements. It helps the customer find your proposal response to a specific RFP requirement. For access by readers, place the matrix in the proposal near the table of contents. An example of the compliance matrix is shown in the model formal proposal later in this chapter.

If the RFP is electronically available in computer files, you can use commercial software to automatically process the files during your solicitation analysis. For example, software can search for key words of interest and can identify the difference between a revised RFP and the RFP it supersedes.

■ *RFP Rule #7: Carefully consider a proposal extension.*

The RFP will dictate a deadline for submitting your proposal. If you believe that you can't produce a competitive proposal within the allotted time, you must then decide to submit the best proposal you can by the deadline, ask that the deadline be extended, or stop proposal work and not bid the RFP. (Note that late proposal submittal isn't offered as an option.) Before you ask the customer for an extension, answer the questions in Table 7–3.

If, after considering the questions in Table 7–3, you believe an extension is necessary, ask for it. It's fair to think that the worst that can happen is that the customer will say no.

If an RFP amendment grants the extension or delays the expected contract award date, ensure that all associated dates are correct in your proposal. For example, the submittal date on a cover or title page may have to be changed, or the time period used for schedules and cost spreadsheets may require revision.

■ *RFP Rule #8: Prepare for RFP amendments.*

An RFP amendment can have minor or extensive impact on the content of your proposal. As described in RFP Rule #4, develop a process for distributing RFP amendments. Try to predict the release of amendments by looking for the customer activity noted in the "Customer intention" portion of Table 7–3. Expecting to receive an amendment can help you identify and commit the resources you'll need to review the RFP changes and then make any corresponding proposal changes. Think twice about delivering the proposal long before it's due. If you submit too early, a late RFP amendment from the customer could cause you to revise, reprint, and redeliver your proposal.

■ *RFP Rule #9: Carefully consider an alternate proposal.*

The RFP may authorize an alternate proposal—that is, a proposal that offers a solution or an approach different from that requested by the RFP. You can submit an alternate proposal as a separate document or offer the alternative as an option in the primary proposal. Check the RFP for submittal instructions. Regardless of where alternative ideas are proposed, if they run counter to the RFP requirements, you need to explain why they are better. Develop and submit an alternate proposal if you wish, but not at the expense of producing a competitive primary proposal. Your first priority should be to provide a proposal responsive to the RFP requirements. Don't offer an alternate proposal if you know the customer doesn't want one.

TABLE 7–3: Factors to Consider before Asking for a Proposal Extension

Factors	Questions and Considerations
Benefit to the competition	*How will the extension help your competition?* If you're given the extension, expect the customer to grant it to all of your competitors. The extension might help your competition more than you.
Competition needs or plans	*Do you think your competition needs an extension?* The customer may be more willing to grant an extension if more than one competitor asks for it. If you think a competitor will ask and you're reluctant to do so, you could wait for that competitor to make the request. If you think you're the only company that needs or wants an extension, you may be forced into making the only request.
Your resources	*Do you have the budget and personnel resources to cover additional proposal preparation time?* If you don't, don't ask for the extension.
Your extension need	*How much time do you really need, and how much time can you realistically expect the customer to give?* The answers to these questions can help you determine the extension length you should request. You might ask for exactly what you need. On the other hand, you might ask for more time than you need, hoping that at least your minimum need will be met. If you don't think the customer will grant you enough time regardless of your request, maybe you shouldn't ask.
Customer response	*How will an extension request be received by a customer that can't or doesn't want to delay the procurement process?* For example, the customer may have little time for delay because the contract must be awarded before the end of the fiscal year to commit budgeted funds or because the requested product or service must be delivered by a certain date, which can't be jeopardized by delaying the solicitation and award process.
Commitment to continue	*After you ask for the extension, will you continue to work on the proposal?* If you don't continue the work, you waste time you may not be able to make up if an extension is approved. It's best to continue proposal work while the customer is considering your extension request. If you can't or won't, an extension may not help you very much.
Request justification	*How will you justify your request to the customer?* Justify your request by explaining how the extension can help you and the customer. Be careful about explaining how your problems forced you to ask for the extension. Be honest, but don't let your explanation to the customer reflect poorly on your company's capability.
Customer intention	*Is there any indication that the customer may initiate a proposal extension?* For example, if questions from the prospective bidders have been submitted to the customer and are still unanswered, the customer may grant an extension to develop and distribute the answers and to allow the competing teams to incorporate the answers in their proposals. Also, if the customer indicates that major changes to the RFP requirements are forthcoming, the RFP amendment with those changes could include an extension.

FORMAL PROPOSAL RULES

Because they usually concern products or services for major projects, formal sales proposals attract a larger, more diverse audience than do letter proposals. Readers often skip around in formal proposals, looking for the sections that most interest them. Expect them rarely to read the proposal from cover to cover at one sitting. Given this proposal complexity and reading pattern, what writing strategies will work for formal proposals? Use the following rules, remembering to apply the writing and graphics guidelines described in Chapters 4 and 5.

■ Formal Proposal Rule #1: Use an "easy-in, easy-out" rhetorical pattern.

Your customers should be able to read your proposal with an "easy-in, easy-out" approach. Promote this rhetorical pattern by using the following features in your writing:

- Short, easy-to-read sections at the beginning—These sections contain overviews targeted at the less technical readers. Included is material such as the title page, transmittal letter, table of contents, list of illustrations, executive summary, and introduction.
- Longer, more detailed sections in the middle—These sections contain the proposal details and are aimed at both nontechnical and technical readers. Middle sections can be grouped into three basic topics: technical, management, and cost.
- One short, easy-to-read section at the end—Provide a concluding section that contains points you want the reader to remember, such as a wrap-up of main benefits. It can make a positive and lasting impression on the decision makers.

■ Formal Proposal Rule #2: Help the reader with introductions and conclusions.

Readers focus on beginnings and endings; therefore, introduction and conclusion sections can be effective communication points in your proposal. Be brief in your introductions and conclusions, realizing that their length depends on the size of the proposal or section they address. As a general rule, use "Introduction" and "Conclusion" headings to identify these descriptions and limit each subsection to a page or two.

Use introductions at the beginning of each proposal volume and at the start of major sections within a volume. In the introduction, briefly address the following in the order listed:

- Purpose—Succinctly state the reason for the proposal.
- Problem/Need —Describe the problem/need the proposal addresses. In a solicited proposal, this section can be short, because the customer has already indicated the need. However, in an unsolicited proposal, the description may be longer because you should explain the problem in detail and justify action to meet the need. If the description becomes lengthy, present it under a separate "Problem" or "Need" heading following the "Introduction."

- Scope—Describe the range of proposed activities, products, and services covered in the proposal, along with any research or other tasks that may have already taken place during preproposal work. Consider the use of a table that lists the features and resulting benefits of your product or service.
- Road map—Describe or list the sections or topics to follow. It serves as a condensed table of contents for readers.

Conclusions are appropriate at the closing of a proposal volume and the end of major sections within a volume. However, a conclusion at the end of a lengthy proposal may not be very effective, because the longer a proposal is, the less likely it will be read in its entirety. In the conclusion, briefly describe the following:

- A summary of the work to be done, and the features of the product or service to be provided
- The main selling points of your product or service, focusing on benefits to the customer
- Your commitment to meeting the customer's needs

The conclusion brings readers back full circle to what was emphasized in the introduction—the major features and benefits of what you are proposing.

Formal Proposal Rule #3: Include an executive summary.

Unless the customer indicates otherwise, always include an executive summary. Because of the potential overlap of information in the proposal introduction and the executive summary, carefully coordinate their content. Follow the executive summary guidelines described in Chapter 6.

Formal Proposal Rule #4: Use a title page that informs, attracts attention, and sells.

Use your title page to convey important information, including the proposal's title and originator, the customer's name, the proposal submittal date, identification of the applicable RFP, and any limits on the use of the proposal content. However, the title page can also be used to attract the reader's attention and support the selling of your proposal by using the following elements:

- Titles—Use titles that reflect the benefits of your product or service. For example, change "Proposal for a New Pension Plan for Jones Engineering" to "Improving the Pension Plan at Jones Engineering," or change "Proposal for Accounting Data Base Program at Acme, Inc." to "A New Accounting Database: Cutting Costs at Acme." Show how your proposal will help the customer, but, comply with your customer's preference when the RFP requires or implies that the RFP and proposal titles are to be the same.
- Graphics—Use a title page graphic that represents the major selling points (themes) of your proposal, because it can strengthen your sales message and attract reader attention. If proposing a nationwide network of sales offices, you could have a map of the United States with stars placed at the location of each

office. If offering a proven and reliable product, you could show a photograph of a person using the product.

■ *Formal Proposal Rule #5: Use a transmittal letter that informs and sells.*

Submit an effective transmittal letter (cover letter) to accompany your formal proposal. Write this letter to provide more than administrative information for your proposal; compose it to sell your proposal as well. Follow the transmittal letter guidelines described in Chapter 6.

■ *Formal Proposal Rule #6: Provide support information near the front of the proposal.*

Provide the following information to help readers understand the content and organization of your formal proposal:

- A table of contents—Accurately list the heading titles and page numbers, generously using white space to make reading easier. Pick a minimum level of heading subordination to show in the table of contents, and comply with this standard. For example, you may have five levels of headings in the body of the proposal but identify headings only to the third level in the table of contents.
- Figure and table lists—Accurately list the figure and table numbers and their associated captions and page numbers. Make two separate lists—one for figures (all graphics except tables) and another for tables. If you identify tables as figures, use only a figures list.
- Acronym lists—Provide a list of acronyms and their definitions. If desired, expand the list to include abbreviations and initialisms. The list often appears in a page format near the beginning of the proposal, but it can be in many forms in different parts of the proposal. For example, it could also be printed on card stock inserted in a pocket of the proposal binder or on a reference tab bound within the proposal. The list could also be in a foldout at the end of the proposal, where it can be unfolded and used while the proposal is being read.
- Compliance matrix—As previously described in RFP Rule #6, this matrix cross-references RFP sections to where they are addressed in the proposal.
- Foreword—This is an optional page that can include administrative and introductory information that may not be in your transmittal letter, title page, or introduction.
- Proposal structure—For a multivolume proposal, a graphic can show how the proposal is organized, indicating where the subject volume fits into the overall proposal structure. The graphic can appear as an organization chart showing the titles and organization of the proposal volumes and their major sections. Highlight a portion of the graphic to show where the relevant volume appears in the proposal structure—similar to the map in the mall that shows the location of its shoppers and stores.

If the customer dictates a proposal page limit, the preceding support information is usually not part of the limitation. However, check the RFP to be sure.

Formal Proposal Rule #7: Explain your technical approach.

> The benefits of the system are derived directly from the customer's fundamental requirements. Not from the whizbang capabilities of systems. So, in a way, it's a sign of gross disrespect to try to sell features without saying what they are good for, for that customer. The message that comes across is that you didn't listen, you didn't analyze the connections between needs and capabilities, you didn't trouble to think about it.
>
> (Svoboda and Godfrey, 1989, p. 120)

The technical section provides the technical details about your product or service—for which you can expect a tough, doubting audience. Use the following techniques to have your readers understand and accept your technical approach:

- Organize the whole and parts deductively. Lead off the main section and every subsection of your technical portions with the single most important point in that part. This pattern makes major points easy to find and understand.
- Follow the appropriate outline. For solicited proposals, take your cue from the customer by arranging technical information in the order dictated by the proposal instructions in the RFP. If the proposal is unsolicited, arrange points in the order in which the needs were identified in your problem/need section. In both cases, repeatedly link what you are proposing to the customer's needs.
- Associate features with benefits. It's not enough to describe the technical features of your proposed product or service. You must also explain the benefits of those features. Also tie the features and benefits to the customer's needs.
- Back up claims with facts. No matter how true the technical claims may be to you, they need support before a skeptical audience. Don't take a "trust me" attitude with the reader; substantiate your claims with proof.
- Use graphics to explain technical details. The use of flow diagrams, schematics, line drawings, and photos can help explain technical details that can be hard to understand in text. Look for opportunities to either replace or supplement text with graphics. Remember that a photo of a product indicates that the product exists and isn't just a conceptual idea.

Formal Proposal Rule #8: Explain your management approach.

You can have a great product or service to offer, but without a sound management approach to provide it, you can fail to meet the customer's needs. To convince the reader that you can effectively manage the project, describe the project schedule, the background and assignment of project staff, and the project management structure and plans. These elements of the management approach include the following:

- Schedule—major project milestones, including delivery dates
- Personnel—qualifications, experience, and project responsibilities of key people assigned to the project

- Structure—organization chart(s) showing the positions, personnel assignments, and reporting levels for project management
- Plans—management processes and tools for monitoring, controlling, and reporting the performance of the project

Table 7–4 provides pointers for describing a complete and credible management approach.

■ *Formal Proposal Rule #9: Explain your risk management approach.*

Even if the customer doesn't ask for it, you should assess the risk associated with the proposed product or service and explain how you will manage and mitigate that risk. What's risk? It's anything that could adversely affect your technical, management, or cost performance in the proposed project. The following are examples of risk:

EXAMPLES

- You propose a new and untested product under development that must operationally perform in compliance with demanding customer specifications. (technical risk)
- To meet the volume demands for a proposed part, your manufacturing division must increase its daily production capacity from 100 parts to 1,000 parts. (technical and management risks)
- Based on your company's poor labor relations in the past, you have a company history of labor strikes, high employee turnover, and an inability to hire qualified people. (management risk)
- A portion of an accelerated project schedule has critical milestones that you must meet on time for the project to succeed. (management risk)
- Customer delay in approving the final design of your product may increase the cost of its production because key product suppliers are planning to raise material costs in the near future. (technical and cost risks)
- Bad weather can prevent project work from occurring on schedule. In addition, the delay can increase labor costs because it idles your workforce. (management and cost risks)

The process of risk analysis is similar to the strategy planning described in Chapter 1. Describe and classify the severity of the risks and explain what you have done or will do to mitigate the risk. Then, after explaining your risk mitigation approach, reclassify the risk level based on the mitigation approach.

If you're completely candid, you typically will find that risk can vary from very high to negligible—that is, there are few, if any, products or services that come with no technical, management, or cost risk. A sound risk management plan can show the customers that you understand their needs and are prepared to meet those needs. An honest approach to identifying and managing risk can give you

TABLE 7–4: Tips for Describing the Management Approach

Project Schedules	
Preparation and credibility	Use activity and delivery schedules to indicate that you have prepared well to provide the product or service. Even a preliminary schedule, subject to later review and change, can make your proposal more credible to the customer.
Approach definition	Draft schedules early in the proposal process, so they can be used to define your technical and management approaches. A project schedule with milestones and specific objectives allows you to link your management plan to the completion of definable tasks.
Management Structure	
Customer interest	Show the levels and positions of management that will most interest your customer and will have the greatest impact on delivering the product or service. For example, the customer may be more interested in your project team organization than the upper-level corporate structure of the president and vice presidents.
Simplicity	Simplify organization charts to avoid the appearance of a complex, top-heavy management structure. If different tiers of management must be shown, consider separate charts for each level.
Management structure change	Show how the project management structure will change as project tasks are completed. As a project progresses, the organizational structure may become more simple or complex.
Team and customer interfaces	Show how supporting organizations, such as teaming companies, subcontractors, and vendors, will be involved in your project management. Also, explain the customer interface with your project management.
Personnel Résumés	
Brevity and consistency	Make résumés brief and easy-to-read in a consistent format. If the customer dictates a format, follow it.
Information highlights	Use bulleted lists, italics, boldfacing, and underlining to highlight important information, so readers can easily scan the résumés for key points. Avoid text-heavy résumés that read like autobiographies.
Project applicability	Have the résumés clearly show what and how personal qualifications, experience, skills, and accomplishments are applicable to meeting customer needs.
Project role	Describe the person's role in the proposed work; a person's title and general job description aren't always enough to explain specific job responsibilities in a proposed project.
Personal touch	To add a more personal touch in the résumés, consider the use of photos and signatures.
Organization chart	In the organization chart(s), identify the position and name of every person who is represented by a résumé.

continued

TABLE 7–4:—*continued*

Personnel Assignment	
Honesty	Be truthful about the availability and commitment of personnel assigned to the project. Don't claim or insinuate that people will be full-time when they won't be. Avoid the "bait and switch" ploy, in which you propose the assignment of project personnel with the intention of replacing them with others if you win the contract.
Time commitment	Note the percentage of the time that personnel will spend on the project, remembering that it's normal for key managers to divide their time among several projects. If you show too many full-time upper-level managers, it may appear that there will be too much management, with a high price tag to match.
Availability limitations	Describe any limiting conditions for personnel assignment. For example, you might guarantee someone's assignment if the contract is awarded within three months of proposal submittal.
Replacement plan	Describe how key personnel will be replaced if they leave the job and what effect, if any, the customer will have in choosing a replacement.
Positions and people	Clearly identify project positions along with the people assigned to those jobs. If possible, propose the assignment of personnel who are employed currently at your company. Avoid the assignment of contingency employees—that is, people who don't currently work for your company but will if the contract is awarded.
Personnel commitment	Have assigned personnel sign commitments of availability on the résumés. Have your highest levels of management, including teaming companies and your major subcontractors and vendors, sign the proposal transmittal letter.
Management Plans	
Processes and tools	Describe processes and tools for monitoring and controlling the technical, schedule, and cost performance of the proposed project. Tools can include technology (hardware and software) used to support management tasks, such as analysis, communication, and reporting.
Project objectives	Explain how management processes and tools will be used to meet project objectives. For example, explain the procedures and standards that will be used to measure project performance and also describe the content, distribution, and use of reports that will allow you and the customer to monitor project status.
Graphics	Use flow and schematic diagrams of your management processes and tools to explain your management approach.
Structure and staff	Associate the management processes and tools with the project personnel who will use them.
Customer involvement	Explain how the customer will be involved in your management plans.
Risk management	Amplify your management approach with a description of the risk management planning addressed in Formal Proposal Rule #9.

an advantage over your competitors who ignore risk issues and pretend that nothing could possibly go wrong.

■ *Formal Proposal Rule #10: Clearly and accurately identify your price.*

Make your proposed costs clear to the reader and link them to value. Also ensure that your cost section accurately reflects the product or service described in the technical and management sections. To coordinate the content of these sections—which are often developed by different people—perform the following:

- Include finance (quoting and pricing) personnel in your early proposal strategy, planning, and kickoff meetings, so they understand the scope and requirements of the proposed product or service.
- Ensure that the people who provide financial information for the cost section understand what's being offered in the proposal. If they also serve as proposal writers related to the quoted product or service, this shouldn't be a problem. If they don't, provide them a current draft copy of the technical and management sections as the price is being developed.
- Review the cost section whenever the technical and management sections are being reviewed. Instruct the review team to look for discrepancies among the sections. Verify that work descriptions, schedules, material listings, labor categories and hours in the technical and management sections match the pricing data in the cost section. Unfortunately, the development of the cost section often lags behind that of other proposal sections. Therefore, it may not be ready for review when other parts of the proposal are.

You may be instructed by the customer to exclude cost figures in your technical and management sections. Even with this limitation, you can *allude* to costs in these sections by citing how your product or service will *save* money in terms of dollars or percentages.

> If your costs are considerably higher than those submitted by others, you will need to be especially careful to explain the additional charges and the need for them. If, on the other hand, your costs are considerably lower than those in the other proposals, you may well need to explain how you can perform the same quality work as others in your industry for a lower cost.
>
> (Bowman and Branchaw, 1992, pp. 167–168)

Avoid making your reader suspicious about your costs; don't bury the price in long paragraphs or write defensively about it. Instead, address the issue of price by using the following techniques:

- Put costs in spreadsheets or formal tables. Cost numbers can be hard to find within paragraphs.

- Emphasize value received for costs. Particularly if your costs are higher than average, mention the main benefits that result from the higher costs.
- Be clear about the cost of add-ons or options. Also, present this information in spreadsheets or formal tables.
- Report costs to the whole dollar. First, it's easier to proofread costs without the decimal point. Second, the figure $123,500 looks smaller than $123,500.00.
- Always total the costs; don't force readers to make their own calculations. When you total the costs, triple-check the accuracy of your figures.

Formal Proposal Rule #11: Explain the applicability of company experience.

A description of a company's work experience can indicate that the firm can provide the proposed product or service. However, to support this contention, you should describe how this experience is applicable to the proposed project. Don't assume that readers will make this connection on their own. Show applicability by explaining how the requirements of your past projects are the same or similar to those of the proposed project. The following are features that can allow you to cite parallels between your past and what you propose for the future:

- Deliverable product or service, including quality and quantity
- Skills and qualifications of project personnel
- Management structure of the project team
- Corporate composition of the project team, including teaming companies and supporting subcontractors and vendors
- Scheduling and reporting requirements for the project
- Location of the customer or the project work
- Technical and management challenges and how they were and will be overcome

Of course, emphasize successful work your company has performed for the customer targeted by your proposal.

An RFP may request specific details about your past experience. It may also dictate a format and page limitation for this response and limit your choice of contracts to those performed during a defined period. During proposal evaluation, the customer may use this information to ask your former customers about your performance. This feedback can have a major impact on the success of your proposal.

You may be forced by the RFP to describe a contract in which your customers weren't satisfied by your company's product or service. If faced with this dilemma, don't ignore the issue. Instead, explain what lessons you learned from that work and what you've done to avoid the problems in the future. You may also know that your target customer isn't pleased with your company's performance in a prior or current project. If so, assume that your proposal readers will be aware of this situation. Be straightforward in your proposal; cite the contract in question and explain what's been done to prevent these problems from happening again.

If your firm has little or no related experience as a business entity, cite the applicable experience your employees had while working for other companies.

■ *Formal Proposal Rule #12: Write with brevity and comply with proposal length limitations.*

Too much detail can drown your message. Therefore, in the main body of your proposal, only provide the text and graphics that your readers need to make their decision. Put the less essential information in well-organized appendices or attachments. Consider the following items as candidates for placement in an appendix or attachment:

- Résumés and organization charts
- Company histories
- Detailed graphics
- Detailed product descriptions or design specifications
- Detailed descriptions of similar projects or contracts
- Detailed procedural descriptions
- Exhaustive cost information
- Marketing brochures

Ensure that this information is understandable, relevant, and organized. Avoid using an appendix or attachment as a dumping ground for extraneous information. If the customer doesn't need the information, don't provide it.

If there are page limitations, plan early to comply by allocating page budgets for all proposal sections—including the appendix or attachment—that count against the page limit. To help you set and meet page budgets, use storyboards and mock-ups to plan the scope of the proposal text and graphics. The following are some other tips to stay within a page allowance:

- Graphics—Place details in your graphics if their placement in text would require more space. Although it's best to have simple graphics, if you have to choose between detailed graphics and not meeting your page limitation, pick the detailed graphics. Foldout pages can be used to display very detailed graphics. However, remember that an RFP may dictate that a foldout counts as two pages, so a foldout may not be the answer to your page limit problem.
- Layout and typography—Select page layout and typographical options that allow for maximum content on the page. The options can include your approach for spacing between text lines; breaking between major sections; spacing after periods (for example, one space instead of two); showing the start of new paragraphs (for example, indention or line spacing); selecting type size and font; and setting the format for columns, page margins, and justification.
- Referral to other proposal parts—You may have to respond to an RFP requirement that was addressed elsewhere in the proposal. If so, avoid duplication and unnecessary use of valuable page space; refer the reader to the graphics or text in the proposal sections that cover the same topic.
- Use of proposal sections with no page limits—As long as you comply with RFP requirements, place details in proposal sections not subject to page limitations. For example, an RFP may allow the submittal of a volume or an appendix that has no page limitation. If it does and you comply with all RFP proposal

instructions, consider placing details in these sections instead of those with page limitations.

AN EXAMPLE AND A CRITIQUE OF A FORMAL PROPOSAL

This section contains a model of a formal sales proposal written to demonstrate the proposal rules in this chapter. *(The proposal is based on a fictitious product and situation.)*

Context of Oil Tanker Proposal. Steve Wilson, a recently promoted maintenance supervisor for Hydrotech, is anxious to obtain a big contract for his firm. His hopes lie with a solicited proposal he has written to Standard Shipping International.

Based on his RFP analysis and the discussions that he and the Hydrotech president have had with Standard Shipping management, Wilson knows the customer wants one firm to perform hull cleaning and maintenance inspections for its fleet of galaxy-class oil tankers. The shipping company now hires various firms for these services. By selecting one contractor, Standard Shipping expects to save money, improve the physical condition of its tankers, and ensure that its fleet is regularly inspected.

Wilson believes several firms will be bidding services in competition with Hydrotech. Although smaller than many of its competitors, Hydrotech uses an innovative and patented system for cleaning hulls. Knowing that none of Hydrotech's competitors can offer this technology, he emphasizes its virtues in the proposal. Wilson had great latitude in how the proposal was written, formatted, and organized because the RFP provided few proposal instructions. He decided to follow the formal proposal guidelines in this chapter and produced the following proposal.

A PROPOSAL FOR HULL CLEANING AND INSPECTION

In response to:

Standard Shipping International RFP #SSI 02-07999
February 1, 2000

The Seasled System: Savings and Increased Efficiency
For Standard Shipping International

Prepared by

Hydrotech Diving and Salvage, Inc.
New Orleans, Louisiana

February 24, 2000

TABLE OF CONTENTS

LIST OF ILLUSTRATIONS

FIGURES

TABLES

COMPLIANCE MATRIX

RFP Requirement	Proposal Section
Show an understanding of the problem.	1. Introduction 1.1 Description of Hull Maintenance Problem
Describe your equipment. Describe your cleaning process.	2.1 Timesaving and Reliable Equipment 2.2 Portability and Mobility That Make Sense 2.3 Team and System Procedures 2.4 Diver Rotation for 24-Hour Coverage 2.5 Completion Times for Responsive Service
Describe the scheduling of your service.	3.1 Scheduling That Saves Money 3.2 Professionals Who Make the Difference
Describe your staff qualifications.	3.3 Other Services Available to You
Submit a firm cost proposal.	4. Reduced and Predictable Costs
Describe your risk management approach.	5. Conclusion

Provides an overview of the problem

Introduces the solution

Focuses on the main benefits

EXECUTIVE SUMMARY

The galaxy-class oil tanker has a large hull surface area below the waterline. Accumulated marine growth can damage the hull and induce excessive drag between the hull and water. The results can be slower ship speeds, increased fuel consumption, and more frequent hull repairs.

The removal of marine growth can be expensive and time-consuming. Hydrotech, however, offers a hull cleaning system that is effective, economical, convenient, and safe. Our patented Seasled system, which has cleaned more than 100 tanker hulls over the past two years, uses a diver-operated, self-propelled cleaning vehicle that can be brought directly to your anchored vessels. The portable system is operated and supported by experienced crew teams carefully selected and trained by our company.

The primary benefits of the Seasled system include:

- Hull cleaning completed in about 80 hours, about half the time required by conventional cleaning systems used by diver teams
- Mobile hull cleaning at times and locations convenient for you, reducing the out-of-service times of your tankers as they are readied for and undergo hull cleanings

Because the Seasled system reduces the cost of each cleaning, you will be able to afford annual hull cleanings and semiannual inspections. This regular attention will increase the average cruising speed and fuel efficiency of your ships. The cost of our services is based on fixed and guaranteed regional rates, allowing you to accurately budget for your cleaning and inspection needs.

Repeats background information for readers who start with the introduction

1. INTRODUCTION

The Hydrotech Seasled system, a new—yet proven—method of hull cleaning, meets your goal of reducing the hull cleaning costs for your galaxy-class oil tankers.

A clean hull increases ship cruising speed and fuel efficiency and reduces the need for hull repair. If the cleaning can be performed quickly, you can minimize the time that the affected ship is out of service and unable to serve your customers.

Our Seasled system is not available from any other firm. It is the best and most economical portable system available to clean tanker hulls and to help you cut hull cleaning expenses.

1.1 Description of Hull Maintenance Problem

Standard Shipping International operates a fleet of eight galaxy-class oil tankers engaged in world-wide transporting of petroleum. Your vessels travel to oceans and ports around the world, where many forms of marine growth collect on ship hulls. Because this tanker class has such a large hull surface area, even light marine growth can cause excessive drag between the hull and water. Furthermore, if marine growth is allowed to accumulate for too long, the hull can deteriorate. Marine growth results in slower ship speed, increased fuel consumption, and more frequent major hull repair.

Links Hydrotech experience to the Seasled system and the proposed services

1.2 Scope

The Seasled system, which became operational two years ago, reflects our 10 years of experience providing hull maintenance and cleaning for tanker fleets around the world. We designed, built, and tested the Seasled system with the assistance of the following maritime consultant firms:

- Yamamoto Shipbuilding Service Consultants, Kure, Japan
- Marine Corrosion Consultants, New Orleans, Louisiana
- Ocean Science Center, Key West, Florida

These firms are internationally known and have supported past Standard Shipping operations.

This proposal explains how our safety-conscious and skilled crews will use the Seasled system to reduce the cleaning time for your tankers. We will also describe the features of our system, the cleaning process, and the crew schedules as well as the backgrounds of our Seasled team members. Then you will find a cost quote for our services.

Prepares the readers to understand the proposal flow and content

1.3 Proposal Format

For ease of reference, the remainder of this proposal is divided into three major sections:

- Seasled: How It Will Work for Standard Shipping—describes the equipment, crew, and cleaning procedures
- Scheduling, Qualifications, and Other Services—provides information about the scheduling of our service, the qualifications of our key personnel and cleaning teams, and other services you may need
- Reduced and Predictable Costs—provides cost information to contract for our cleaning and inspection services

2

2. SEASLED: HOW IT WILL WORK FOR STANDARD SHIPPING

This section describes the Seasled equipment and procedures used for cleaning the galaxy-class tanker hull.

2.1 Timesaving and Reliable Equipment

The Seasled system is a portable, diver-operated device capable of cleaning 500 square feet of hull surface per hour. **Figure 1** shows a three-dimensional view of the Seasled vehicle and its major components.

Describes the major system components and how they work

The underside of the vehicle is equipped with three major components:

- Rollers—Rubber rollers allow the vehicle to move along the hull as the diver looks for marine growth and positions the vehicle for hull cleaning. With the vehicle close to the hull, the diver can closely inspect the hull and quickly move the vehicle to areas of interest.
- Air nozzles—Powerful nozzles use suction to attach the vehicle to the hull during inspection and cleaning operations. When the diver finds a hull area that requires special attention, the vehicle is positioned next to the hull and then attached with suction nozzles. The nozzles can also be used to blow air to help loosen and remove marine growth after it has been scraped.
- Scrapers—Hardened rotary scrapers, next to the rollers, loosen and remove marine growth from the hull. They can be used while the vehicle moves along the hull or while the suction nozzles attach the vehicle to the hull.

An umbilical cable connects the vehicle to a portable control module aboard ship. Each vehicle is controlled by a separate control module. Through the cable, the control module provides the vehicle with electric power and nozzle air. Electricity is used to propel the vehicle, to power vehicle electronics, and to move the scrapers. Air is used for the suction and blowing functions of the nozzles. The cable, heavily wrapped and insulated to protect its internal feed lines, also provides a digital communication link between the diver and module tender operator. The link allows the two people to pass operational and safety information to better coordinate the cleaning activity.

A control panel allows the diver to control vehicle propulsion and operate the scrapers and nozzles. The panel also contains a communication panel for sending and receiving digital messages between the diver and tender operator. A steering wheel provides the diver with pinpoint control of vehicle turns. To protect against personal injury, a windshield buffers the diver from loosened marine growth as the vehicle moves through the water.

Cites the proven capabilities of the system

The Hydrotech-developed Seasled system has successfully cleaned more than 100 tanker hulls for our customers. We have never had a system failure that prevented us from meeting a hull cleaning commitment in the contracted time.

Before our divers began using the Seasled vehicle, portable removal of marine growth required them to hand carry and operate equipment during the dives. This manual approach is still used by most companies that perform portable hull cleaning.

Based on our operational results, we have found that the Seasled system not only does a better cleaning job, but it also completes the work about 50 percent faster than the manual diver approach. Surveys conducted with all our customers show that our system has cut their hull cleaning expenses by 40 percent and increased the annual at-sea availability of their ships by 10 percent.

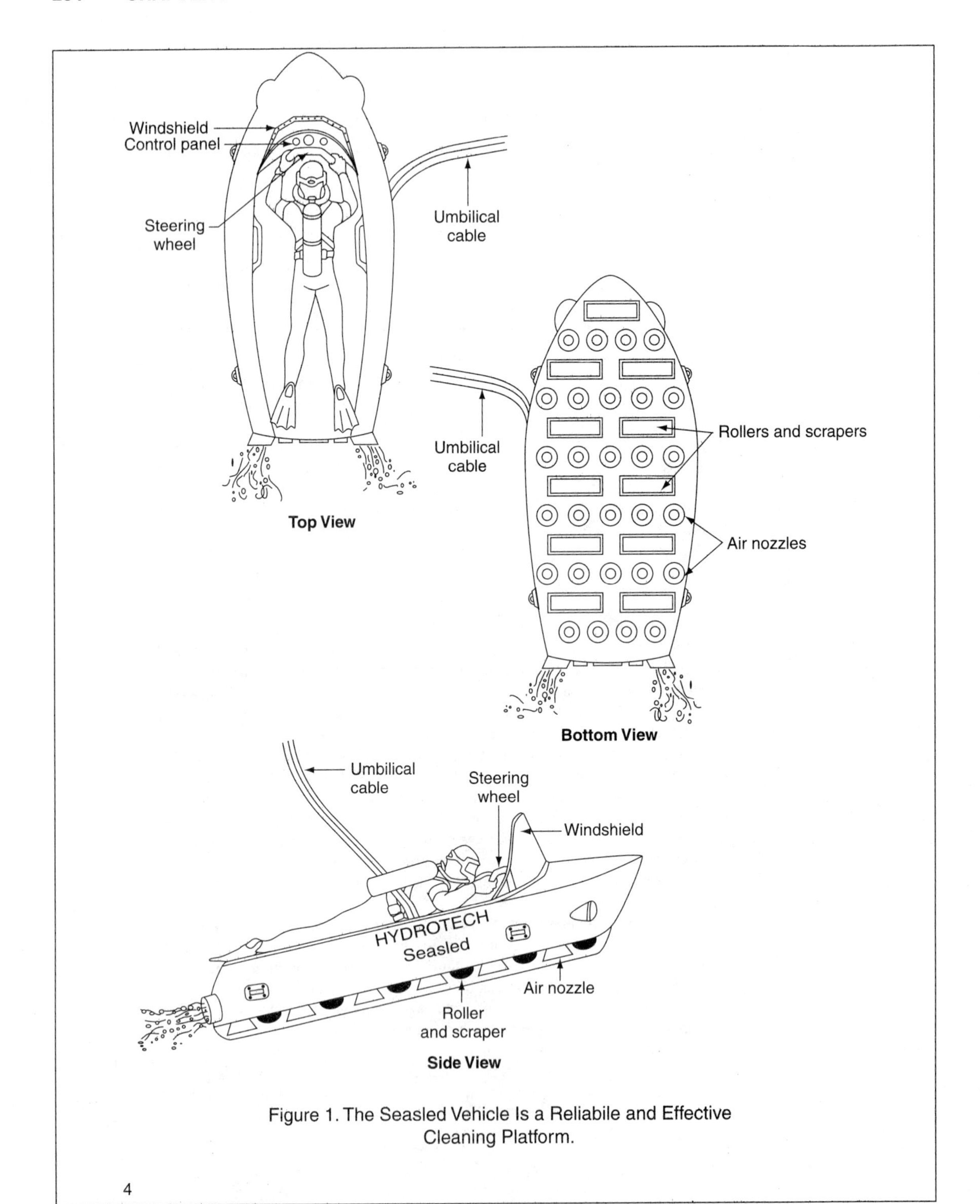

Figure 1. The Seasled Vehicle Is a Reliabile and Effective Cleaning Platform.

4

2.2 Portability and Mobility That Make Sense

In the past, tanker operators avoided portable hull cleaning because of its inefficiency and time requirements. They often preferred to take vessels to deepwater ports where expensive pier space was used to perform scheduled hull maintenance. Tankers might have to travel hundreds of miles to reach these ports. With the portability and mobility of the Seasled system, you can now transform any scheduled or unscheduled layover of your tankers from lost shipping time to convenient maintenance time.

Emphasizes two key benefits of the system

The Seasled equipment is easily transferred from a small craft to a tanker deck, using the cargo lifting davits already on board. It can also be delivered by helicopter on ships with a helo pad. Furthermore, the hull cleaning can be conducted while the ship is anchored in open sea or in port.

By using our Seasled system, you can now have the mobility, speed, efficiency, and convenience of a hull cleaning system that can be transported to your vessels at any location.

2.3 Team and System Procedures

Our Seasled divers and support crews consist of highly trained and experienced professionals. A typical cleaning team includes the following:

- Two experienced diving supervisors to coordinate the placement and operation of the Seasled vehicles and divers
- Six certified commercial divers to control and operate two Seasled vehicles
- Four vocationally trained tenders to coordinate the transport, storage, and logistical support of the vehicle platforms and diver equipment
- Two qualified equipment technicians to service, maintain, and repair the vehicle platforms and diver equipment

Clearly describes how the team operates the system

Trained as a team, these personnel transport and operate the system. Once aboard your ship at the desired location, our team can have the Seasled system in place and operable within 10 hours.

Figure 2 shows the functional process of cleaning a hull. Major hull cleaning tasks include the following:

- Two vehicles are lowered into the water, where they are boarded by the divers.
- The divers steer and position the vehicles beneath the ship's waterline. The vehicles, placed on the port and starboard sides of the ship, are operated from bow to stern to clean the hull.
- Each diver and vehicle are monitored and assisted from topside by a tender operator in the control module.
- Two standby divers are on deck and ready to assist the working divers in case of an emergency.
- An equipment technician monitors the operation with the tender operators and is available to perform support module or vehicle repairs.

During the process, the divers are equipped with independent life-support systems and standard diving and safety equipment. They are also coordinated by a diving supervisor.

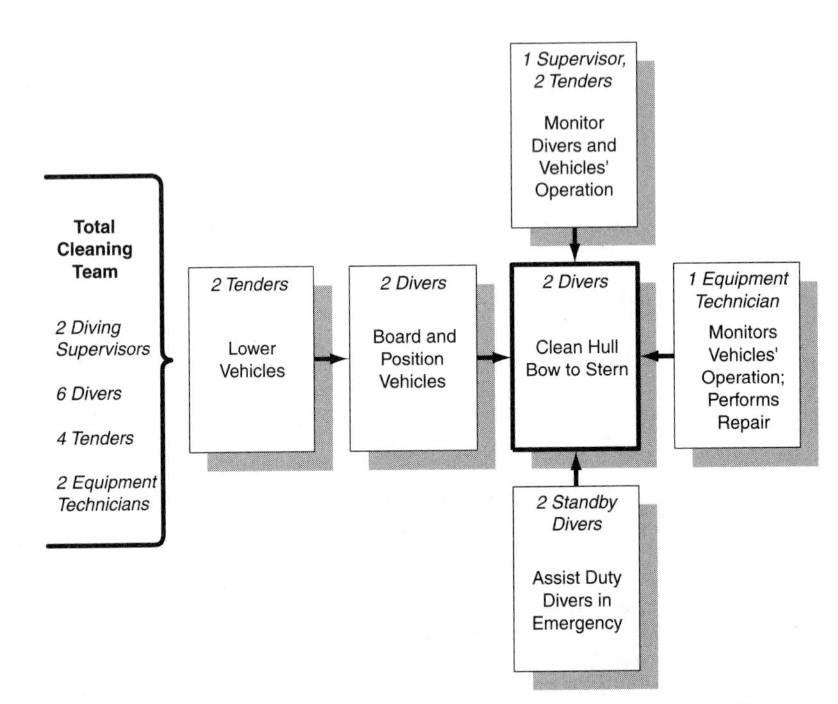

Figure 2. Hull Cleaning Is Performed by an Experienced and Skilled Seasled Team.

2.4 Diver Rotation for 24-Hour Coverage

Our divers work 12-hour shifts around the clock with a diving supervisor on duty at all times. The divers are assigned to the following schedule:

- Oncoming divers act as standby divers for the four hours immediately preceding the start of their diving duty.
- On-duty divers work four hours and are then relieved.
- Relieved divers are off duty for 12 hours following their dive duty.
- After the off-duty period, divers can be placed on call for four hours before beginning standby duty.

The diver rotation ensures the safety and productivity of our divers and minimizes the time required to complete the cleaning job.

2.5 Completion Times for Responsive Service

Table 1 lists the typical time required to complete major system support and operational tasks for cleaning the galaxy-class tanker hull. The average time to complete the process is 80 hours, or about 80 hours faster than the portable

Amplifies a description of the process

TABLE 1: Hull Cleaning Can Be Completed Typically within 80 Hours, Much Faster Than Conventional Diving Methods.

Task	Average Completion Task Description	Time
System setup	Position crew, equipment, supplies, and materials on the ship; provide an operational briefing to the tanker crew; and set up and test the support modules	10 hrs.
Hull cleaning	Inspect and clean the hull	65 hrs.
System removal	Pack system equipment, supplies, and materials and remove them and Hydrotech personnel from the ship	5 hrs.
Total time:		**80 hrs.**

cleaning method used by conventional diver teams. The times are based on documented in-service results and reflect the average cleaning times for an empty galaxy-class tanker with a draft of about 30 feet and 60,000 square feet of hull surface below the waterline.

3. SCHEDULING, QUALIFICATIONS, AND OTHER SERVICES

This section describes our service schedule and the qualifications and roles of our team personnel. It also summarizes other Hydrotech services that can support your tanker fleet.

3.1 Scheduling That Saves Money

Refers to customer needs and how they'll be met

As you have indicated in the RFP and our previous discussion, Standard Shipping wants each ship to have annual cleanings and semiannual inspections to prevent the heavy marine growth on your fleet hulls. The thicker the encrustation, the harder it is to remove. Therefore, you need a regular maintenance schedule to keep your ships operating at peak performance and to help us complete each job on time.

Realizing that your ships run on irregular schedules due to the constantly changing requirements of your customers, we will provide the following scheduling features to meet your needs:

- Maintain a record of each ship's annual cleaning and semiannual inspections and contact you to arrange a periodic rendezvous for hull cleaning
- If you need unscheduled work, mobilize a Seasled team to arrive at your ship's location within 72 hours after receiving your request

Our scheduled and unscheduled services will help you maintain a seaworthy and profitable tanker fleet.

Cites the experience and qualifications of Hydrotech personnel

3.2 Professionals Who Make the Difference

Since we began our business 10 years ago, Hydrotech has recruited professionals experienced in all phases of diving, ship construction, and repair. Our personnel have the education and experience necessary to provide safe, efficient, and professional services to our customers. Our top management, representing more than 100 years of maritime and diving operations, include the following personnel:

- William Bailey, Jr., president and chief executive officer, Hydrotech—formerly the Commander, Naval Undersea Systems Center, USN (Retired)
- Joseph Smith, senior vice president, Hydrotech—formerly the Commander, Naval Ship Repair Facility, USN (Retired)
- John Delong, Diving Personnel Superintendent, Hydrotech—formerly a master diver, Naval Diving and Salvage Facility, USN (Retired)
- Samuel Johnson, operations officer, Hydrotech—formerly the Commander, Naval Undersea Construction Center, USN (Retired)

All Seasled team members are carefully selected by our personnel department, under the supervision of Mr. Delong. The following are the qualification and training requirements for specific positions:

- Diving supervisors—selected from our senior divers or former USN master divers, who must also qualify as Seasled divers
- Divers—selected from our senior tenders or former USN first class divers, who must then pass our Seasled handling course
- Tenders—selected from graduates of certified commercial diving schools, who must complete our in-house training program
- Equipment technicians—selected from trade-school graduates, who must have at least one year of experience in our equipment repair shops

Our demanding selection and promotion policies are necessary for us to maintain our exceptional productivity and safety records. Our personnel safety performance is the best in the hull cleaning industry. No Hydrotech employee has ever been injured on the job.

Summarizes other services the customer may need

3.3 Other Services Available to You

In addition to the hull cleaning service, Hydrotech offers many maintenance and emergency services, including

- Removal of obstructions from propellers and shafts
- Sea suction and discharge clearing
- Recovery of lost equipment
- Emergency damage-control repairs
- Assistance to vessels that have run aground
- Ship salvage operations

Unexpected problems such as these are all too common in the shipping industry. Response time often determines whether schedules will be met, whether equipment can be recovered, or even if lives are saved. At Hydrotech, we will respond to your unexpected needs with speed and dedication. You will receive the highest priority in an emergency.

8

Clearly shows costs linked to value

4. REDUCED AND PREDICTABLE COSTS

The Seasled system introduces increased efficiency and, thus, reduced costs to hull cleaning projects. The charge for our services varies by location, due to transportation costs. For your convenience, we have assigned flat rates for five world-wide regions:

- Region 1—Southeastern United States (includes Alabama, Florida, Georgia, Louisiana, Mississippi, South Carolina, and Texas)
- Region 2—Northern quadrant of the Western Hemisphere (except that area designated Southeastern United States)
- Region 3—Southern quadrant of the Western Hemisphere
- Region 4—Northern quadrant of the Eastern Hemisphere
- Region 5—Southern quadrant of the Eastern Hemisphere

Figure 3 depicts these five regions and shows where the six Hydrotech offices are located.

Table 2 lists the charges for scheduled and unscheduled hull cleaning and scheduled inspections. These per-ship costs are guaranteed for two years after contract signing. If we are unable to mobilize our team on your ship within 72 hours for your unscheduled cleaning need, we will respond as soon as possible and discount our quick response charge by 10 percent. The fixed nature of these costs will allow you to budget for and control your hull cleaning and inspection expenses.

These rates apply only to empty galaxy-class tankers with a draft of about 30 feet. Special rates can be quoted for loaded tankers on an individual basis. Rates are also available for our other services, including maintenance and emergency support. As indicated in the RFP, Standard Shipping will provide room and board for our teams.

TABLE 2: Our Cleaning and Inspection Rates Are Fixed by Region and Guaranteed for Two Years.

Task	Region 1	Region 2	Region 3	Region 4	Region 5
Scheduled cleanings	$50,000	$60,000	$75,000	$80,000	$80,000
Unscheduled cleaning (72-hour response)	$60,000	$70,000	$85,000	$90,000	$90,000
Semiannual inspections	$10,000	$12,000	$15,000	$16,000	$16,000

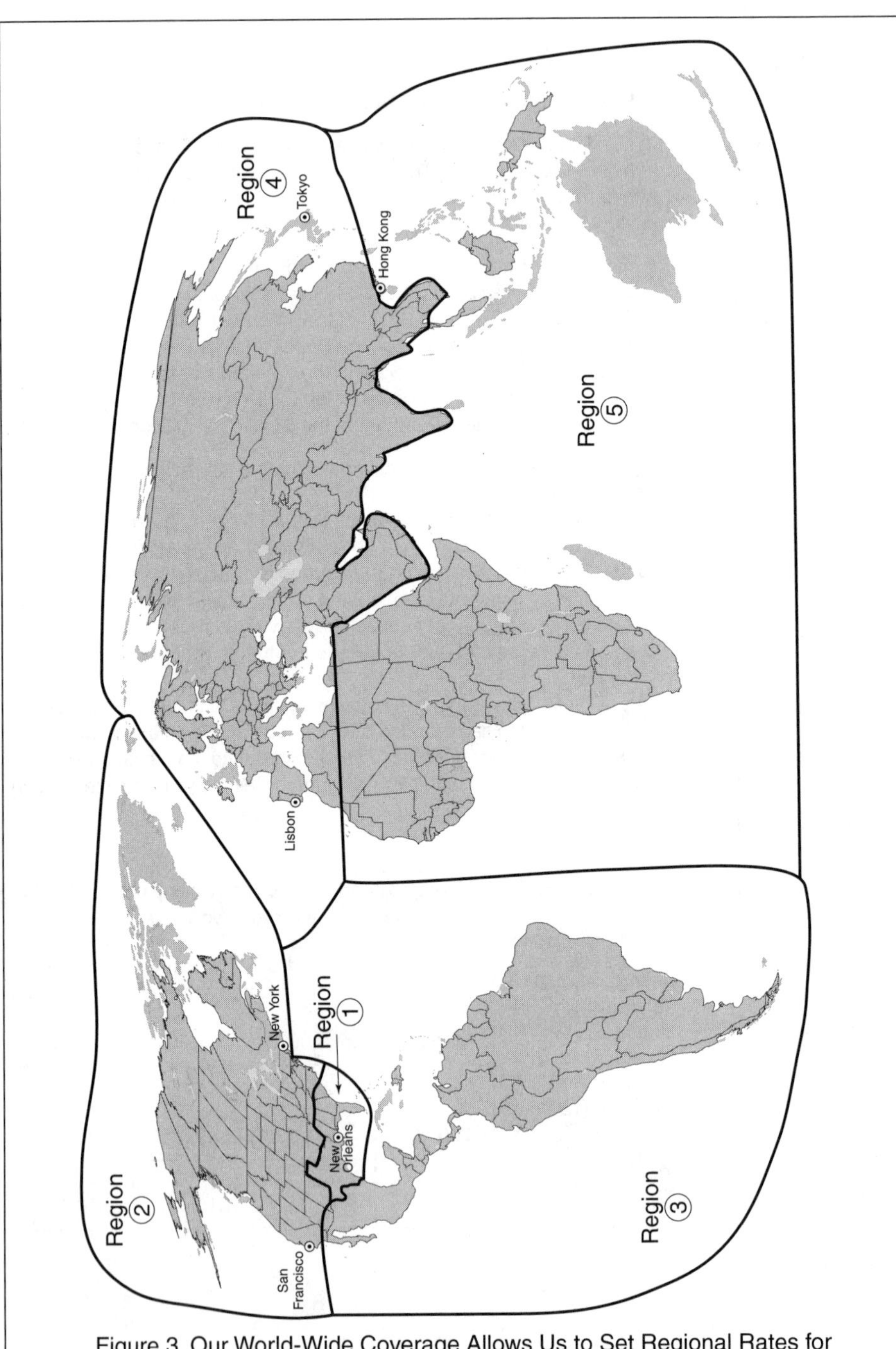

Figure 3. Our World-Wide Coverage Allows Us to Set Regional Rates for Cleaning and Inspection.

5. CONCLUSION

Summarizes the needs and solutions

The hull cleaning of galaxy-class tankers can be expensive and time-consuming while using conventional diving teams or fixed cleaning services in deepwater ports. The Seasled system uses a diver-operated scrubber vehicle, a portable system that can clean the bottom of an empty galaxy-class oil tanker in about three days. It will quickly clean your tankers anchored in the open sea or in port, while reducing your cleaning costs, transport times, and mooring expenses. Fixed and guaranteed rates, based on the location of the cleaning site, will allow you to accurately budget for cleaning and inspection costs.

Addresses how risk will be reduced

Our teams will rendezvous with your ships at times and locations that fit your schedule and can turn lost time into money-saving cleaning time. To keep your ships operating at maximum efficiency, we recommend annual cleanings and semiannual inspections for your fleet. If you require unscheduled hull cleaning, we will provide a 72-hour response after receiving your request. In addition to our hull cleaning service, we can perform unscheduled and emergency ship repair, salvage, cleaning, and recovery work.

Hire Hydrotech with the assurance that your hull cleaning will be performed safely and economically in a timely fashion. The Seasled system is a low-risk approach for meeting all of your safety, operational, and cost requirements:

- The risk of injury to our crews and your liability for our employees working on and around your ships are reduced by the documented safety and reliability of the system and the experience and training of our Seasled crew teams.
- The Seasled system can be used in seas with swells of 8 to 10 feet, allowing hull cleaning in many sea conditions.
- Because of our world-wide logistic and transportation network, we can quickly mobilize and support our Seasled teams to serve your fleet in various ports and open sea anchorage areas.
- As demonstrated over the past two years, the Seasled system cleans hulls 50 percent faster than conventional diving methods and can be expected to reduce your hull cleaning expenses while increasing the at-sea availability of your fleet.

More details about our Seasled system are available. If desired, Hydrotech will provide a free demonstration of the system. We are committed to meeting your hull cleaning and inspection needs. Our system can affect your "bottom line" in two ways: Hydrotech will keep your hull bottoms clean and also increase your profits.

A Formal Proposal Critique. As the proposal manager, Wilson has written a responsive proposal for Standard Shipping. It addresses the customer issues of expense and scheduling and clearly explains the features and benefits of the Seasled system. Here are the ways that the proposal complied with the 12 rules for writing formal proposals:

- *Rule #1*—The beginning and ending sections of the proposal are easy to read, with more detailed information contained in the longer middle sections.
- *Rule #2*—An "Introduction" section makes clear the purpose of the proposal, the extent of any preproposal work, the scope of information in the proposal, and the format of the rest of the document. It also includes a brief description of the problems caused by marine growth on the tanker hulls. (A longer problem statement would require a separate proposal section after the "Introduction.") A "Conclusion" section brings the reader back to the major benefits of the Seasled system.
- *Rule #3*—The "Executive Summary" highlights the major selling points of the proposal and serves as a lead-in to the proposal.
- *Rule #4*—The main title reflects the RFP title. However, a subtitle is added by Wilson to emphasize increased efficiency. The use of a customer logo and a graphic of the globe on the page recognizes the worldwide presence of Standard Shipping. The graphics make the page more attractive and interesting. Wilson obtained permission from Standard Shipping to use its logo in the proposal.
- *Rule #5*—The transmittal letter, shown as an example in Chapter 6, uses lay language to explain how Hydrotech can solve Standard Shipping's problems. By briefly describing the Seasled system and its main advantages, the letter supports the selling of the Seasled service and engages the reader's interest rather than wasting the letter on empty clichés and generalizations. Also, it ends with a promise to call the customer to discuss the offer of a Seasled demonstration. The letter supports the proposal administratively and topically. The letter was written and signed by the Hydrotech president to show the commitment of the highest level of company management. However, Wilson helped the president prepare the letter by telling him the major proposal themes.
- *Rule #6*—White space is used effectively in the table of contents. The sections and subsections are evenly spaced over the page. Aligned indention permits readers to scan chapter subheadings with ease. The support information includes a list of figures and tables. All headings and captions are listed as they appear in the proposal. The accuracy of all page numbers was carefully checked. A compliance matrix is included to show where RFP requirements are addressed in the proposal.
- *Rule #7*—The technical sections cover the major features and benefits of the Seasled system and its operation. Graphics summarize the technical and operational details. If they had been available, it would have been preferable to use photos of the Seasled vehicle. The photos would have emphasized the existence of an operational and available system, an impression that may not be supported as well with drawings. Benefit claims are substantiated by the results of actual system use. Benefit-oriented headings attract readers to important topics.

- *Rule #8*—The management sections address the scheduling of cleanings and inspections and the qualifications and assignments of management and staff. Detailed organization charts and résumés of key Hydrotech personnel weren't included, because the RFP didn't request them and Wilson believed that the customer wasn't interested in those details.
- *Rule #9*—The "Conclusion" section explains how the Seasled approach will minimize safety, cost, and operational risks—concerns that the customer would probably share.
- *Rule #10*—The cost section clearly reflects the Seasled service described in the technical and management portions of the proposal. It links cost to customer value and displays rates in an easy-to-read table. The cost table was carefully checked for accuracy.
- *Rule #11*—The proposal cites corporate and personnel experience and stresses the experience Hydrotech has gained serving customers with the Seasled system. Documented results substantiate the benefits of using the system.
- *Rule #12*—Although the RFP imposed no page limitations for the proposal, Wilson decided to write a brief but informative proposal. However, in the "Conclusion" section, Wilson offers to provide more details and proposes to conduct a system demonstration. The liberal use of tables, figures, and bulleted lists helps to limit proposal length and enhance its readability. Action captions for the graphics support the "selling" of the Seasled service.

SUMMARY

Formal proposals are more than five pages long, excluding appendixes and attachments. Typically bound with a cover, they can be defined as a sales proposal, grant proposal, or planning proposal. Although their content, organization, and format vary, formal sales proposals often contain the following information:

1. Purpose
2. Technical approach
3. Management approach
4. Related experience/past performance
5. Cost

A formal proposal can be solicited (requested by the customer) or unsolicited (not requested by the customer). If solicited, it's developed in response to an oral or written request from the customer. A request for proposal (RFP) is a written solicitation. If responding to an RFP, you must meet all RFP requirements and instructions to produce a competitive proposal. The following rules will help you review, use, respond to, and possibly influence RFPs:

1. Participate in draft RFP reviews.
2. Submit comments, questions, and recommendations.
3. Assign an RFP coordinator.
4. Control RFP configuration and distribution.

5. Participate in bidders' meetings.
6. Use requirements and compliance matrixes to be responsive.
7. Carefully consider a proposal extension.
8. Prepare for RFP amendments.
9. Carefully consider an alternate proposal.

Use the following rules to write competitive and responsive formal proposals:

1. Use an "easy-in, easy-out" rhetorical pattern.
2. Help the reader with introductions and conclusions.
3. Include an executive summary.
4. Use a title page that informs, attracts attention, and sells.
5. Use a transmittal letter that informs and sells.
6. Provide support information near the front of the proposal.
7. Explain your technical approach.
8. Explain your management approach.
9. Explain your risk management approach.
10. Clearly and accurately identify your price.
11. Explain the applicability of company experience.
12. Write with brevity and comply with proposal length limitations.

EXERCISES

1. Formal Sales Proposal to City or Town: Individual Assignment

Write a formal proposal in which you suggest a change in the services offered by your local city or town. Target an audience that might conceivably receive such a proposal. Do local research and then perform the audience analysis described in Chapter 4. Assume that you are writing this proposal as a representative of an interested community group. The proposal is unsolicited, so you must provide a convincing statement of need. Fortunately, you've been able to meet once with your main reader, who agreed to read and consider the proposal you plan to submit.

The following are some ideas for a proposed change:

- Changing an area's traffic pattern (proposed by a subdivision representative to the city's mayor and traffic engineer)
- Adding a new outdoor sports field (proposed by the representative of several softball leagues to the recreation director and city council)
- Rescheduling city trash pickups (proposed by representatives of several civic clubs to the sanitation director)

2. Formal Sales Proposal to a College: Individual Assignment

Write a formal sales proposal to propose a change in the operating procedures, personnel, physical plant, curricula, or any other feature of a college or university.

Target an audience that might conceivably receive such a proposal. After completing campus research, perform the audience analysis described in Chapter 4. Assume that the college has formally solicited the proposal from outside organizations. You are writing as a representative of one of these groups.

The following are some ideas for a proposed change:

- Adding a video-monitor system for informing the campus community about college activities (proposed by an electrical contracting firm)
- Improving the appearance of all or part of the school's landscaping (proposed by a local landscaping firm that has done similar projects for other colleges)
- Adding a degree program to the curriculum (proposed by a professional society representing companies that want to hire students with the new degree)

3. *Formal Sales Proposal to a Company: Individual Assignment*

Write a formal sales proposal in which you, as representative of one firm, propose to sell a product or service to another firm. Select a product or service with which you are reasonably familiar, on the basis of your work experience, research, or interests. (Don't use proprietary information about the product or service.)

Assume that your proposal is solicited and has been preceded by many letters, phone calls, and meetings between you and the customer representative. You're convinced the proposal will get a fair reading, but you also know that several other firms will compete for the contract.

4. *Formal Sales Proposal to a College: Team Assignment*

Use the RFP exercise in Enclosure (2) to produce and evaluate a team-written proposal. The exercise also includes instruction for giving an oral presentation in support of the proposal.

5. Commerce Business Daily (CBD): *In-Class Group Discussion*

As a group, obtain copies of at least one week's worth of the *CBD*. (You may be able to obtain *CBD* copies from libraries or businesses in your area. *CBD* information from the U.S. government can also be obtained via the Internet.) Discuss the organization and content of the publication, noting the variety of products and services being requested by various U.S. government agencies. Find examples announcing the (1) pending release of an RFP and (2) the solicitation for A/E services via the SF255. Then discuss the content of these announcements.

6. *U.S. Government RFPs: In-Class Group Discussion*

As a group, obtain copies of recent RFPs released by at least three different U.S. government agencies. (You may be able to obtain printed RFPs from companies or federal agencies in your area. RFPs can also be obtained from U.S. government web sites on the Internet.) Compare the content and organization of the RFPs, with particular attention to the requirements in Sections A, C, L, and M.

7. Grant Proposal Planning: Individual or Group Assignment

Find a local library that contains grant proposal information from the Foundation Center or a similar source. Become familiar with the library's reference material that identifies potential funding for nonprofit organizations. With this reference material, find foundation sources that could potentially fund the following non-profit activities in your local area:

- AIDS awareness and education for teenagers
- Job training and placement for homeless men and women
- Recreation programs for children with physical disabilities
- Computer training for elementary school students

8 Informal Proposals

OBJECTIVES

- **Learn the content and use of two informal proposals: letter and memo proposals**
- **Learn rules for developing letter and memo proposals**
- **Examine models of letter and memo proposals**
- **Perform exercises using chapter guidelines**

Informal proposals are shorter and less elaborate than the formal proposals described in Chapter 7. However, like formal proposals, they can be written for an external or in-house audience. An informal proposal for an external reader is a letter proposal; for an internal reader, it's a memo proposal. This chapter describes when and how to develop letter and memo proposals—both of which share many of the same developmental rules.

LETTER PROPOSALS

How do letter proposals compare with the formal sales proposal? First, both propose that some action be taken. A letter proposal, however, uses a letter format rather than the multisection, bound format of a formal proposal and has no more than five pages (excluding attachments). The letter proposal is appropriate in the following situations:

- The customer specifically asks for a letter proposal.
- The customer asks for an informal proposal and provides no guidance about its format and organization.
- In the absence of customer instructions for proposal format and organization, you believe the customer would want a letter proposal or you think the small size of your proposed project or sale warrants a letter proposal rather than a formal proposal.

A letter proposal consists of two parts: the letter and any attachment(s). The letter contains the most important information for decision makers; an attachment contains supporting detail.

Rules for Developing Letter Proposals

> In certain respects, a short proposal is more challenging to write than a long proposal. Each sentence must carry a heavy information load.
>
> (Miner and Griffith, 1993, p. 67)

As you would do for other types of proposals, begin planning your letter proposal by considering customer needs. Then write your letter proposal with the brevity of a letter but with the thoroughness of a formal sales proposal. In addition to the proposal writing and graphics guidelines in Chapters 4 and 5, respectively, use customer instructions and the following rules to develop your letter proposals.

■ *Letter Proposal Rule #1: Use an "easy-in, easy-out" rhetorical pattern.*

As previously described for formal proposals in Chapter 7, use the "easy-in, easy-out" approach for your letter proposals. Make the beginning and end short and easy to read; the middle can be a bit longer and more detailed. Although the body paragraphs should average five to eight lines long, varying their length gives visual interest. For example, use short paragraphs—even as short as one sentence—to draw attention to major benefits and other main points; use longer paragraphs to supply less important information.

■ *Letter Proposal Rule #2: Use the 3-Cs pattern to gain and keep interest.*

The 3-Cs (capture, convince, and control) pattern works for letter proposals just as it does for sales letters. This pattern allows you to

- Capture reader interest in the first paragraph or two
- Convince the reader to accept your proposal with strong middle paragraphs in the main body of the proposal
- Control the proposal process by developing an ending that leads naturally to the next step

The next five steps will support the making of a 3-Cs pattern.

■ *Letter Proposal Rule #3: Capture interest in the opening section.*

> *Avoid the clichés of business writing*—the hackneyed openings and closing that your reader has probably seen a hundred times before. In particular, avoid clichéd openings, such as, "I would like to take this opportunity to thank you for considering the enclosed . . . blah, blah, blah." Get to the point.
>
> (Sant, 1992, p. 112)

Attract reader interest with an opening section that uses the type of techniques listed in Table 8–1. If you prefer, this section can include a heading—perhaps "Introductory Summary" or "Overview." Although a heading might seem a bit formal near the opening of a letter, it can draw readers into the proposal.

■ *Letter Proposal Rule #4: Use a deductive writing pattern.*

Readers want important information up front. They don't want to wade through a lot of verbiage. Writing deductively simply means starting with the most important information and then following up with the facts and details that support

TABLE 8–1: Opening Section Techniques for Capturing Reader Interest

Technique	Example
Refer briefly to what prompted the proposal.	"Your RFP requests proposals for completing the new westend parking lot at Sharp Mall."
Mention the customer's main need.	"Your greatest concern, understandably, is that the lot be completed before the three new stores open."
Give a capsule version of what you propose.	"This proposal describes a four-phase plan of construction."
Emphasize a major benefit.	"Completion can be scheduled for May 15, well within your deadline."
Mention the proposal's contents.	"Described below are the four phases of construction, our qualifications for assuming the project, a construction schedule, and a lump-sum cost estimate."

your initial main point. Use this deductive pattern in the following parts of the letter proposal:

- Each main section of the proposal—for example, "The four construction phases comprise removing the mounds of refuse, grading the surface, applying new pavement surface, and painting the space lines."
- Most paragraphs within sections—for example, "Removing refuse from the site during Stage 1 will require about three days and the use of two subcontractors."

An exception to this rule can occur in the cost section, in which you may choose to discuss the value of a product or service before mentioning cost. A preliminary description of benefits can buffer the impact of your price and encourage readers to think first about the value they'll get for the money.

■ *Letter Proposal Rule #5: Use headings that attract attention.*

Letter Proposal Rule #3 notes that a heading is optional for the opening section of the proposal. However, make headings a required part for the sections that follow the opening section in the letter. Use them to engage readers' interest and to make it easier for readers to find proposal topics. Use the following techniques for developing letter proposal headings:

- Ask a question ("What Will Option #2 Cost?").
- Stress a benefit ("Using Subcontractors to Cut Costs").
- Mention a customer need or concern ("Goal: Completing the Job on Schedule").

■ *Letter Proposal Rule #6: Use a conclusion section that emphasizes the main benefit(s) of the proposal.*

Readers remember first what they read last. You can control reader response to your letter proposal by highlighting the proposal's main benefit(s) near the end of the letter. Use a "Conclusion" heading to separate the final wrap-up from the paragraphs that precede it and to signal the reader that important information follows.

■ *Letter Proposal Rule #7: Place an approval block after the closing.*

Make it easy for customers to accept your proposal. Include a few lines such as "approved by" and "date of approval," along with a request for them to return a signed copy to you. With this approach, the customer can accept your proposal by signing and dating the approval block and won't have to write a formal letter of response.

■ *Letter Proposal Rule #8: Use attachment material to make the letter short.*

Use well-labeled attachments to reduce the length of the letters. Suppose, for example, that the modest size of your proposed project suggests the need for a letter proposal, but your outline indicates you'll need about nine pages of text. To reduce the size of the letter, you could place details and supporting information in one or more attachments.

■ *Letter Proposal Rule #9: Use graphics when needed.*

Use graphics in the letter and attachments to visually present information to your readers. To accomplish this,

- Place simple and essential graphics in the body of the letter—for example, a table listing project options and related costs.
- Place complex and less important graphics in the attachments.

■ *Letter Proposal Rule #10: Edit carefully.*

Don't let the informality of a letter proposal lull you into thinking it can be less than flawless. Edit carefully to avoid errors in mechanics, style, and grammar. The following are key rules for editing a letter proposal:

- Triple-check all cost figures.
- Triple-check the spelling of personal names. (Call the reader's company for a correct spelling of the reader's name and title, if you have doubts.)
- Check the format and wording of all headings.
- Check that all attachments are referred to and included.

Chapter 10 has detailed guidelines for editing.

An Example and a Critique of a Letter Proposal

The following example provides a model for your letter proposals.

The Context of a Geotechnical Services Proposal. James Mason has worked as a geotechnical engineer for Tangley, Inc., for about eight years. Most of Tangley's work is providing consulting services for other larger firms involved in massive construction projects. A variety of engineers, technicians, and support people work for Tangley. For larger jobs, Tangley usually receives RFPs and then assembles proposal teams within the firm to answer the RFP. For smaller jobs (under about $75,000), the firm usually receives a verbal request, by phone or at a meeting, along with a written project description. Even though it does much of its work for repeat customers, Tangley is almost always in the position of competing with three or four firms that do similar work.

Mason had a meeting with Arnold Phretz, a construction manager with the large firm of Midwest Constructors. Midwest has the contract to build a mammoth shopping mall in the fast-growing northern suburbs of Yotango, Ohio. In his hour-long meeting with Mason and Susan Lanter (another Tangley engineer), Phretz explained that he wanted a proposal for a detailed geotechnical investigation of the mall site. After evaluating this information during the project, Tangley would be expected by Midwest to recommend an appropriate foundation design and safe construction methods appropriate for the soils and foundation of the site.

Mason knew that the value of this relatively small job would be in the range of $20,000 to $30,000; thus he planned to respond with a short letter proposal. From his previous experience with Midwest, he assumed the proposal would be read by Phretz, then by a construction expert assigned to the mall project, and finally by an accountant and perhaps a few managers in the firm's main office. Mason then wrote the letter proposal shown in the example. *(The following letter proposal is fictitious. In addition, the technical content is not necessarily accurate.)*

TANGLEY INC.
CONSULTING ENGINEERS
307 Ganner Street
Lanston, Ohio 44060
(214) 424-1237

February 15, 2000

Mr. Arnold T. Phretz
Midwest Constructors
1212 Fannin Street
Columbus, Ohio 43216

Dear Arnold:

Emphasizes customer's concern

Susan and I enjoyed meeting with you last week and learning about Midwest's new mall project. Given the varied soils that have been found near the mall's proposed site, we understand your concern that a thorough geotechnical study must precede construction. As you mentioned, your design team needs detailed geotechnical information for the site.

Introductory Summary

This proposal suggests a three-phase program for us to provide the needed geotechnical information. We can easily complete the work by the May 18, 2000 deadline, assuming no unseasonable weather delays occur. To show how our investigation can meet your objectives, these proposal sections follow:

Gives a road map to the sections that follow and stresses the ability to finish the job on time

- Project Description
- Conducting a Thorough Site Study
- Testing and Analyzing Samples
- Submitting Comprehensive and Timely Reports
- Keeping Costs Low
- Conclusion

Leads off with a project description to ensure that the writer and reader have the same understanding

Project Description

From the documents you gave us at our meeting, we understand that you plan to construct a regional shopping mall at a 30-acre site near the intersection of Route 36 and Lambert Road, about 15 miles north of the Yotango Interstate 285 Loop. The proposed development consists of the following:

- Five two-story structures for stores
- Interconnecting walkways
- Two stand-alone, one-story warehouses

Although your plan notes that exact building orientations have not yet been determined, it does show the approximate locations and column loads.

Uses heading to identify study feature

Places details in an attachment rather than cluttering the text

Refers to the customer's own statement about sampling needs

Uses lead-in summary statement to transition to more details

Mentions the new lab and a key benefit of the lab for this project

Explains how previous experience and new software will benefit the customer

Uses heading to identify features of the reports

Uses bullets to highlight contents of the project report

Conducting a Thorough Site Survey
During this phase of the project, we will drill 15 borings to collect undisturbed samples from the site.

Locations. Deeper borings will be located in areas to be occupied by buildings; the shallower borings will occur in areas to be covered by parking lots or walkways. The selected areas will ensure a thorough analysis of the site area. (See Table 1 in Attachment A for a complete list of borings and their depths.)

Drilling Methods. In line with current practice, we will use the dry auger method from the surface down to about 10 feet deep and the wet-rotary method below that. Of course, if water is encountered at any time during the dry auger process, we will stop drilling for a least 10 minutes to get information and to estimate the degree of seepage into shallow excavations.

Sampling Process. You indicated in our meeting that you wanted samples to be obtained at 2-foot intervals up to a depth of 10 feet, and at 5-foot intervals from 10 feet to the final boring depth. Two types of equipment from our Yotango office will be used to collect samples:

- A three-inch, thin-walled tube for cohesive soils
- A two-inch, split-barrel tube for granular soils

After removing samples from the sampling devices, our technicians will examine materials and visually classify all samples. Then representative portions will be sent to our lab in Yotango to promptly begin testing and analysis.

Testing and Analyzing Samples
Our study will include a complete testing and analysis program to develop appropriate foundation recommendations.

Lab Testing. Our full-service geotechnical lab in Yotango, just completed in December, will be used to analyze all samples taken from the mall site. Its nearness to the site will minimize the time to transfer samples to the lab and to coordinate any follow-up soil sampling. (See Table 2 in Attachment A for a list of the proposed tests and their purposes.)

Data Analysis. Once the lab tests yield results, data from the mall site will be analyzed in our Yotango office adjacent to the lab with the help of our patented software program developed for northern Ohio soils. Supervising this analysis will be Dr. Harold Moore, our in-house consultant on all boring programs. Dr. Moore also supervised the analysis on the Hingley Mall project you completed in Oberlin last year. The application of this software, directed by Dr. Moore, will ensure a successful analysis approach tailored to your Ohio mall site.

Submitting Comprehensive and Timely Reports
We will give you a complete geotechnical report on all phases of the project. It will include three main types of information for your review by your design and construction experts:

- Details about the field work, lab tests, and engineering analysis
- Recommendations about suitable types of foundations and approximate bearing pressures
- Suggestions for any cautionary steps that should be taken during construction of the foundations

In addition, during the project we will keep you informed of our field, lab, and analysis progress with daily oral reports to your project representative.

Indicates a benefit in the heading

Keeping Costs Low

Our project will minimize drilling and testing costs without sacrificing the quality of the resulting geotechnical recommendations. Our project approach will provide the following features and benefits:

Associates features and benefits with cost—prepares reader for fixed fee bid by noting approaches to lower costs

- Full-service testing equipment at our new Yotango lab, which is within 20 miles of the mall site, ensures close coordination among our on-site, lab, and analysis personnel, and reduces time needed to transfer samples and drilling data to our lab.
- Truck-mounted drilling equipment near the boring locations allows us to respond quickly to scheduled drilling times and to limit the amount of equipment that will be stored on the mall site.
- Experience over the past three years working on four similar shopping center locations, all with soils like those at the proposed site, makes us familiar with the equipment and processes needed to perform the required drilling, testing, and analysis on your site, and provides us with experienced personnel to apply this knowledge to your project.
- A new computerized data-analysis program improves the accuracy of test results, especially for northern Ohio soils, while greatly reducing the number of engineering hours needed to analyze the lab test data.

Specifies terms and conditions of the bid price, rather than "burying" them in an attachment

These benefits mean that we can complete this project for a fixed price of $20,000. This low fee assumes that our work begins no later April 15, boring locations are staked by your survey team, and bad weather does not limit access to boring locations by our truck-mounted equipment.

Conclusion

Returns to the main concern of the customer—getting quality work done on time

Your needs are clear: if mall design and construction are to proceed on schedule, you must have quality geotechnical recommendations from the mall site by your May 18 deadline. As we have done in the past, Tangley Inc. stands ready to complete the project on time for a fair price.

Retains control by promising to call

Arnold, I'll give you a call later this week to see if you want any additional information about how we can meet your needs. If you decide to hire Tangley for the proposed geotechnical services, please indicate acceptance of this proposal by signing a copy of this letter in the space below and returning it to me.

Sincerely,

James Mason

James Mason, Engineer Supervisor
Tangley Inc.

Indicates that there is one attachment

nb
Attachment (A)

ACCEPTED by Midwest Constructors:

Makes it easy for customer to accept the proposal

By: ____________________

Title: ____________________

Date: ____________________

ATTACHMENT A: Borings and Testing Program

The 15 undisturbed sample borings will occur at the depths shown in Table 1.

TABLE 1: Borings

Boring Site #	Projected Use for Each Site	Boring Depth
1–5	Retail building	30 feet
6–7	Warehouse and distribution center	25 feet
8–12	Pedestrian walkway	15 feet
13–15	Parking area	10 feet

The lab testing program will classify soils and test the soil strengths. For these purposes, we will use the tests listed in Table 2.

TABLE 2: Testing Program

Test Category	Specific Test
Classification tests	• Liquid limit testing • Plastic limit testing • Water contents testing • Sieve analysis through #200 Sieve
Strength tests	• Unconfirmed triaxial compression • Unconsolidated-undrained triaxial compression

A Letter Proposal Critique. After having discussed the project with Phretz during the preproposal meeting, Mason felt that Midwest Constructors wanted to award the job to Tangley. Regardless, Mason had the good business sense to submit a well-written, responsive proposal to secure the deal, just as if he had tough competition for the contract.

The example proposal satisfies the 10 rules for developing letter proposals. The following comments describe how the proposal complies with these rules:

- *Rule #1*—The opening and closing sections of the proposal, and the paragraphs within them, are short and easy to read. The middle sections are somewhat longer, but the use of lists and subheadings makes the main body of the proposal flow well.
- *Rule #2*—The "capture, convince, control" strategy is evident in the proposal.
- *Rule #3*—Mason captures interest by mentioning the customer's primary need—a thorough study done by May 18—and by giving a clear road map for the rest of the proposal, including a list of sections.
- *Rule #4*—Most main body sections operate deductively by beginning with the most important point, followed by supporting details. The "Testing and Analyzing

Samples" section, for example, first notes that the project includes a complete testing and lab analysis program; subsections on both program components follow. The "Keeping Costs Low" section is an exception to the deductive pattern: Mason intentionally precedes the cost estimate with a persuasive list of benefits the customer will receive.

- *Rule #5*—The proposal uses engaging headings that stress a benefit or feature to the customer—for example, "Keeping Costs Low" and "Submitting Comprehensive and Timely Reports."
- *Rule #6*—The "Conclusion" section leads back to the major benefit for the customer—quality recommendations by the required deadline at a fair price.
- *Rule #7*—For convenience and to encourage a quick response, an approval block follows the letter's closing. Thus, the customer doesn't have to write a formal acceptance letter.
- *Rules #8 and #9*—More details about the proposed borings and tests are placed in an attachment. If he had thought it necessary, Mason could have provided even more details in the attachment text and tables.
- *Rule #10*—The proposal has been carefully edited.

MEMO PROPOSALS

You suddenly come up with a great way to save your company money. You talk to your boss, who also likes the idea and decides right then to make the change, and that's that. But, if your idea goes beyond the routine, your manager may want some time to consider it and ask you to "put it in writing"—that is, to write an in-house proposal.

Proposing changes in the way your organization operates is usually directed to your boss and other internal decision makers. Called memo proposals, these in-house proposals are in the format of an informal memorandum. Like letter proposals, memo proposals are presented with few trappings. The memo is no more than five pages long. Attachment material is provided for support and to limit the length of the memo.

In-house proposals can also be written in a formal format, as bound documents that contain many separate sections and pages. However, expect your managers to prefer that you reserve this formal effort for proposals to a customer, not to them.

Memo proposals can effectively recommend that changes be made in an organization. Topics vary as widely as do companies and the creative impulses of their employees, but some examples are listed in Table 8–2.

Even if your proposal is rejected, writing a good memo proposal can lead your manager to see you as a valued employee—one who shows initiative and writes well.

TABLE 8–2: Reasons for Using a Memo Proposal

Purpose	Details
Adding a new product line	A software engineer proposes developing a new software product to help customers manage their proposals.
Changing an existing product line	A diaper firm's marketing representative proposes changing the placement of adhesive tape on infant diapers.
Changing company procedures	An automobile assembly line supervisor proposes changing the method for inspecting spot welds.
Changing company equipment	A regional office manager proposes shifting from an individual computer workstation system to a networked workstation system.
Changing company personnel	A regional sales manager proposes closing one office and opening another because of changing sales patterns.

Rules for Writing Memo Proposals

> The success of intracompany proposals requires taking company political, social, and tactical facts of life and business into consideration as well as the inherent requirements of proposals.
>
> (Meador, 1991, p. 135)

A memo proposal derives its format from the internal memorandum; that is, it uses the conventional "date/to/from/subject" beginning. Importantly, it should reflect sensitivity to the political climate of an organization. In addition to the proposal writing and graphics guidelines in Chapters 4 and 5, respectively, use the following 10 guidelines to develop memo proposals.

■ *Memo Proposal Rule #1: Use an extended form of the "easy-in, easy-out" rhetorical pattern.*

Write your memo proposals with the extended form of the "easy-in, easy-out" rhetorical pattern used for letter proposals and formal proposals. For this pattern, use the following features:

- Short, easy-to-understand paragraphs (two to five lines each) at the beginning, immediately after the "date/to/from/subject" block
- Longer paragraphs (four to ten lines each) in the middle sections

- Short paragraphs (two to five lines each) at the end to give the reader visual incentive to read the last section

This pattern allows readers to move quickly into and out of the proposal and provides them the option of reading more detail in the middle of the document.

■ *Memo Proposal Rule #2: Use the 3-Cs pattern to gain and keep interest.*

As it can for letter proposals, the 3-Cs pattern can help draw readers into your memo proposal by doing the following:

- Capturing their interest in the first few paragraphs
- Convincing them of your ideas in the body
- Controlling their response and the entire proposal process with the final paragraphs

The next five steps will support the making of a 3-Cs pattern.

■ *Memo Proposal Rule #3: Capture interest with a subject line and a summary section.*

There are two ways to engage the reader's attention early in the memo proposal.

First, briefly mention a proposal feature and benefit in the subject line—for example, use "Subject: Reducing Mail Costs by Changing Carriers," instead of "Subject: Proposed Change in Mail Carriers." Make the subject line sell your idea, but keep it as brief as possible—preferably under 10 words. This brevity can capture the reader's attention and make it easier for the reader or writer to choose a subject with which to file the proposal.

Second, start the memo body with a proposal summary, labeled "Introductory Summary" or "Overview." Similar to the executive summary of a formal proposal, this section presents a capsule version of the memo proposal's most important points. It can include a brief reference to any previous discussions, the need for the proposed change, your response to this need, and a list of the main sections to follow.

The heading "Introductory Summary" or "Overview" appears immediately after the "date/to/from/subject" block, and the associated section should be as brief as possible—no more than two or three short paragraphs—because it's only a summary.

■ *Memo Proposal Rule #4: Convince readers by including a section about need.*

Because most in-house proposals are unsolicited, you often must persuade readers that there's a real need for the change you suggest. Use the following guidelines for developing a need section:

- Place the section right after the summary section. (Avoid the temptation to go into detail about the need in the summary, because the summary may become too long and the need section too short.)
- Use a heading that evokes interest. (Asking "What's Wrong with the Present Mail System?" creates interest and makes a better heading than "Need for a New Mail System.")

- Start the section with a summary answer to the question "Why is change needed?"
- Use the rest of the section to provide supporting details.

Memo Proposal Rule #5: Convince readers by following the need section with details of the proposed change.

Between the need section and the conclusion, sell ideas to readers as you would do in the technical, management, and cost sections of a formal sales proposal. Be sure to answer questions such as these:

- Exactly what needs to be done?
- What will be the main benefits?
- How will success be evaluated?
- How long will the project take?
- What people, facilities, and materials will be required?
- How much will it cost?

As in other parts of the memo proposal, make this section as readable as possible by putting the most important points at the beginning of paragraphs and sections and by using headings and subheadings that engage interest.

Memo Proposal Rule #6: Convince readers by thinking and writing politically.

The term "company politics" can have negative implications, but think politically to increase the chance of your in-house proposal being successful. Consider the feelings, positions, and company background of each reader. The following are recommendations for writing the proposal to reflect the politics of your organization:

- View the proposal within the context of the entire organization, because you can expect that your readers who are managers for that organization will read it that way.
- Be particularly tactful if any of the readers helped set up the system you are proposing to change.
- Anticipate reader objections and provide counterarguments.
- Give appropriate credit to your readers. If any part of your proposed idea came from one of them, make that point clear in the proposal. Readers are much more likely to support a proposal if they share ownership in it. When your proposal comes to be viewed as "their idea," it can be easier to get their support for the change.

Memo Proposal Rule #7: Write a conclusion section that emphasizes the main benefit(s) of the proposal.

As previously noted, readers remember first what they read last. The conclusion is your chance to leave an important parting thought in the reader's mind, and lead the reader to action. Use this section to restate the most attractive benefit(s) of your memo proposal. You might want to stress

- Significant technical features and benefits of the idea—if your readers are most interested in how your idea will work
- Main features and benefits of the schedule or your personnel—if time and people are main concerns of your readers
- Costs—if you know that the bottom line and cost savings are what readers most want to know

■ *Memo Proposal Rule #8: Use attachment material to make the memo short.*

Keep the memo length within five pages. Detailed or technical supporting information can be placed in a well-labeled attachment, to reserve the memo only for information that decision makers will need.

■ *Memo Proposal Rule #9: Use graphics when needed.*

Use graphics to express yourself visually. Keep the following guidelines in mind:

- Incorporate graphics into the memo text when they are simple and needed for immediate reinforcement.
- Place graphics in attachments when their main purpose is to supply supporting reference material.
- Keep all graphics as simple as possible.

■ *Memo Proposal Rule #10: Edit carefully.*

Edit the proposal as carefully as you would for a formal or letter proposal. The following are key editing rules for memo proposals:

- Keep sentences short and simple.
- Don't use terms the main reader won't understand.
- Double-check the correctness of all cost figures, graphics, and headings.
- Ask a friend or colleague to read the memo for clarity, appropriateness of tone, and mechanics.

As previously noted, Chapter 10 has details about editing.

An Example and a Critique of a Memo Proposal

The following example provides a model for your memo proposals.

The Context of Sales Training Proposal. Sanford Mertz works as training director for Mudd Engineers, a medium-sized civil engineering firm with 15 main departments. Mudd recently put all department managers through a rigorous and costly sales training course conducted on company premises by an outside expert.

To evaluate the quality of this training effort, Mertz also attended the course. Although he liked it, he knew that it was attended by only a handful of staff members who actually needed sales training. Reduced training funds precluded giving the course to a wide audience, so Mertz developed a sales training plan that would use only in-house talent instead of expensive outside consultants. He arranged to meet with the corporate manager of engineering, Tom Baker, who must approve all training efforts. Baker responded favorably to Mertz's plan and asked for a memo proposal to consider. Mertz then wrote the memo proposal shown in the example. *(The following memo proposal is fictional. In addition, the technical content is not necessarily accurate.)*

INTEROFFICE MEMORANDUM

MUDD ENGINEERS

DATE: January 21, 2000
TO: Tom Baker, Manager of Engineering
FROM: Sanford Mertz, Training Director
SUBJECT: Effective In-house Training for Sales Skills

Subject line indicates key proposal feature

Introductory Summary

Mentions need and solution and refers to previous discussion with Baker

Using our discussion last week as a starting point, I've devised a preliminary plan to provide sales training throughout the firm. The goal is to deliver this training quickly and economically, while keeping much of the effectiveness of more expensive alternatives.

This proposal notes the need for continued training and then proposes one-day courses entitled Sales Techniques, Proposal Writing, and Sales Presentations. As Figure 1 shows, these three courses can work together to strengthen our sales training program.

Only general features of the courses are provided below. Details can be developed after you've provided your views about the training approach and content.

Sales Training and Mudd's Future

Identifies company need—department managers want more sales training

Last month's three-day sales training course was highly rated by all 15 department managers in attendance. Judging by the written comments from their course evaluations, it's clear they think we need much more sales training if we're to stay competitive in our traditional markets. Also, they were pleased that the course helped them come up with at least one new marketing opportunity for each department.

Introduces the main problem—company can't afford supervisor training

Most encouraging of all, the managers said they would require all supervisors reporting to them to attend similar courses—if we make such courses available. That brings us to the problem we discussed the other day, Tom. This year the company can't afford the training expense of running all supervisors through the same training that department managers received.

Identifies another company need—proposal and presentation training

In addition, some department managers have suggested we go one step further to provide not only a general sales course but also more specific training in two areas: proposal writing and sales presentations. Based on my preliminary research, if taught by outside consultants, I think these training programs would be as costly as the sales training.

Leads into the next three sections

Our problem, then, is that we have an immediate need for training without the funds to hire outside trainers. The rest of this proposal outlines my solution to this problem: we can economically provide the needed sales, proposal, and presentation training by relying on several in-house experts.

Indicates training course and purpose

Mentions main point—the condensed course

Uses bullets to highlight important features

Sales Techniques Course: Learning to Sell Technical Services

As you know, I also attended the three-day sales course last month taught by Hank Wills from Sales Design Ltd. My first suggestion is that we condense that material and develop our own one-day version of the course. Here are some points that make this an attractive option:

- A member of my staff, John Rail, successfully taught a similar course three years ago for his former employer. I feel confident that he could develop and teach this new course.
- Sales Design Ltd. has given us permission to reproduce any written materials used in last month's course. The course can be tested on a "pilot" group and, if successful, repeated many times.

Returns to a main theme—exposing many employees to training

This one-day course wouldn't be as thorough as the course taught by Sales Design Ltd. But it would provide adequate time to cover highlights of effective sales techniques and to provide several hours for role model exercises. The short seminar's main benefit is that we can expose many employees to this training, while removing them from billable projects for only one day.

Indicates training course and purpose

Restates need for proposal training

Emphasizes qualifications of an in-house resource

Proposal Writing Course: Learning to Write Persuasively

As noted above, many department managers wrote in their course evaluations that we need related courses in proposal writing. I suggest that the second part of our sales training program be a one-day course for proposal writing.

Three years ago our editor, Jane Edwards, taught several in-house technical writing courses. The courses, which received high marks from our engineers who attended the training, covered reports, instructions, and proposals. With our new emphasis on sales, Jane has told me that she would be glad to develop a seminar specifically focused on sales proposals. Jane's work on this seminar also might help her complete two related projects recently suggested by members of the engineering staff:

Uses bullets to list other possible benefits

- A brief set of format and content guidelines for writing sales proposals
- A standard section or "boilerplate" document for sales proposals, which could be downloaded from a computer database and then tailored for each new sales proposal

Another advantage of this course is that it would provide a forum for both managers and engineers to agree on approaches to proposal writing. Perhaps we could then reduce the time managers now spend on editing the proposal writing of our engineers.

Indicates training course and purpose

Substantiates need based on the feedback from response forms

Sales Presentations Course: Learning to Listen and Speak

Managers attending the sales course also expressed interest in oral presentations training. A one-day course on this topic would be an excellent way to complete the sales training effort.

The response forms from a recent customer survey show that our customers have three main complaints about presentations by our engineers and marketing staff:

- Too often, presenters don't seem to highlight the real benefits of the proposal. Instead, they talk too long about our firm's history and its successful projects.
- The presentations often exceed the customer's requested time limit.
- Graphics are often not used enough, and when they are they're often ineffective.

In addition to improved presentations to customers, this training could also upgrade our in-house presentations.

As we have seen, Roberta Brown and Charles Claxton of your engineering staff always give excellent presentations. I've talked with them and they're willing to develop and conduct a sales presentations course. Presentation practice in this course could be supported by the new video camera and monitor system we purchased earlier this year.

Returns to cost theme, noting in-house resources

Shows good politics—mentions willingness to support Baker, whatever his decision

Retains control, saying he'll call Baker to arrange a meeting

Conclusion

Our outside consultant taught an excellent sales course last month, but we need to go further. Let's maintain our momentum by giving three one-day courses for sales skills, proposal writing, and sales presentations—taught by in-house experts, rather than expensive consultants.

This memo proposes only a preliminary outline for the training program. I'm looking forward to meeting with you to get more of your ideas on the subject. Whatever direction you decide to take, Tom, my staff is ready to work with you to meet your training needs. As you mentioned last week, at this point in the firm's history, there's no higher training priority than sales skills.

I'll phone you next week to set up another meeting. In the meantime, please call me if you have any questions.

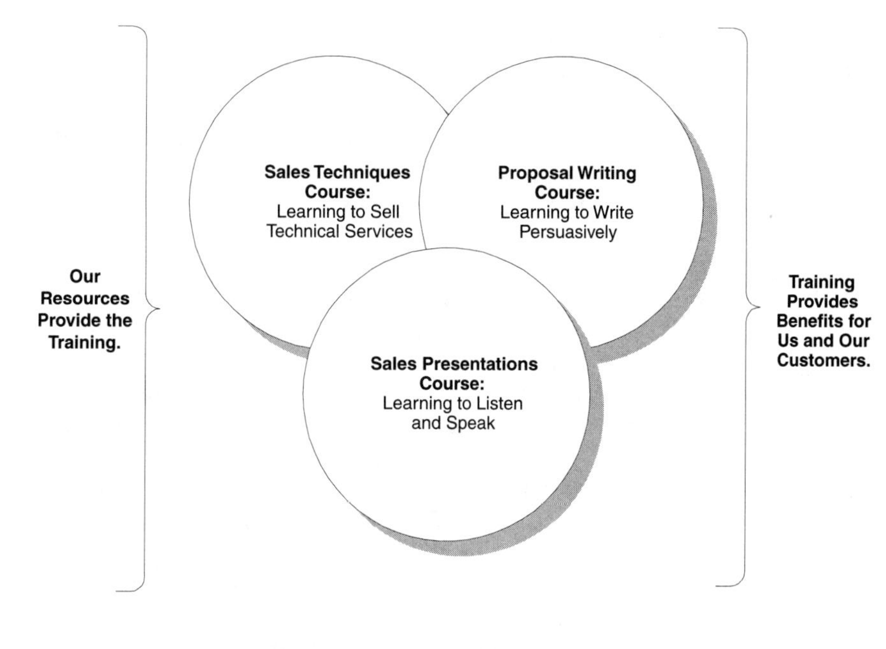

Figure 1
Courses will complement each other, helping us to strengthen our sales training and better serve our customers.

A Memo Proposal Critique. In this proposal, Mertz made an offer he thought that Baker would find hard to refuse. After all, the three training courses will use in-house talent to teach three needed courses. Despite Mudd's need for sales training, however, Mertz wrote a persuasive proposal knowing the proposal would not be accepted automatically. The training program will take many employees away from their jobs for a considerable time, which they would otherwise be billing to customers for engineering work. Also, Baker needed assurance that the courses could be taught successfully with Mudd's own trainers. It would be a false economy to save the expense of an outside consultant only to have in-house trainers do mediocre work.

This example proposal satisfies the 10 rules for developing memo proposals. The following comments describe how the proposal complies with these rules:

- *Rule #1*—The first and last sections of the proposal are short and easy to read, providing the overviews needed by busy readers. The middle of the proposal, however, is more expansive, giving readers details for evaluating the proposed training program.
- *Rule #2*—Mertz adopts a "capture, convince, control" strategy to move Baker toward accepting the proposal.
- *Rule #3*—The subject line first captures attention by stating a goal the reader strongly desires—effective sales skills training. By adding "In-house" to the line, Mertz sets the scene for his points about economy and quality. The "Introductory Summary" section also captures attention by linking the proposal with a previous discussion between writer and reader, outlining the firm's need for sales training, and listing the courses described in the proposal. Mertz notes the proposal's intended limitation: it provides only general features, not details about the three courses.
- *Rule #4*—Mertz realizes that his proposal must thoroughly describe the need for the training program. Because this proposal is unsolicited, Mertz makes no assumptions about Baker's views even though the two have discussed this need. In the "Sales Training and Mudd's Future" section, the seed is planted. Mertz shows that comments from the company's managers themselves have made clear the need for this sales training in all three subjects. The section ends with an effective lead-in to the rest of the proposal. Mertz admits that the company cannot afford outside consultants to provide needed sales training and then provides a transition to the three proposed training courses.
- *Rule #5*—The three main body sections show how the proposed courses can solve the company's sales training problem. Mertz emphasizes features and benefits, such as condensed one-day courses; experienced in-house trainers to keep training costs down; exposure of many employees to the training; relevance of a proposals course to current projects to establish proposal guidelines and standard sections; and improved presentations for in-house and external audiences.
- *Rule #6*—It's clear that Mertz is sensitive to in-house politics and knows how to write diplomatically. First, he mentions that the need for this training has been expressed by the employees themselves, not by trainers. It would appear self-

serving if the need were based on the advice of the trainers. Second, he suggests that some of Baker's own people, who are excellent speakers, could teach the oral presentations course. Third, he emphasizes that this proposal offers only a preliminary plan as a starting point for discussion. Baker, the reader, will have the opportunity to influence the training program.

- *Rule #7*—The "Conclusion" section restates Mertz's most important points: the need to maintain the momentum begun in the sales course last month, the savings made possible by using in-house experts, the training staff's interest in helping with the project, and Mertz's eagerness to meet with Baker soon.
- *Rules #8 and #9*—Mertz presents this proposal as only a "preliminary outline," as he indicates in the "Conclusion" section. There's no need for detailed attachments. If the proposal had been more exhaustive, however, it might have used attachments to provide course descriptions, training media and equipment, instructors' résumés, lists of participants, and schedules for training sessions. Instead of an attachment at the end of the proposal, Mertz includes a simple illustration, which he uses to emphasize the value of a three-pronged effort to meet Mudd's need for sales training.
- *Rule #10*—Before submitting the proposal, Mertz asked several colleagues in his department for comments on content, style, and mechanics; then he proofread the revised copy three times.

SUMMARY

Informal sales proposals are shorter than formal sales proposals and have a less rigid format. Also called letter (for external use) and memo (for internal or in-house use) proposals, they are no more than five pages long, excluding attachments. Attachment material can be added to provide supporting details and limit the length of the letter or memo. Informal proposals are appropriate for making less complex offers of products or services, and when you know or think that your reader wants an informal proposal.

Along with customer instruction, use the following rules to develop letter proposals:

1. Use an "easy-in, easy-out" rhetorical pattern.
2. Use the 3-Cs pattern to gain and keep interest.
3. Capture interest in the opening section.
4. Use a deductive writing pattern.
5. Use headings that attract attention.
6. Use a conclusion section that emphasizes the main benefit(s) of the proposal.
7. Place an approval block after the closing.
8. Use attachment material to make the letter short.
9. Use graphics when needed.
10. Edit carefully.

Use in-house proposals to persuade decision makers to make changes within your own organization. They can help your management identify you as a valued employee with initiative. In-house proposals can be either formal or informal, but expect your in-house audience to prefer them informal. Use the following rules to develop memo proposals:

1. Use an extended form of the "easy-in, easy-out" rhetorical pattern.
2. Use the 3-Cs pattern to gain and keep interest.
3. Capture interest with a subject line and a summary section.
4. Convince readers by including a section about need.
5. Convince readers by following the need section with details of the proposed change.
6. Convince readers by thinking and writing politically.
7. Write a conclusion section that emphasizes the main benefit(s) of the proposal.
8. Use attachment material to make the memo short.
9. Use graphics when needed.
10. Edit carefully.

EXERCISES

1. Letter Proposal to a School: Real Content

If you're a student at a college or university, select one small-scale change you might propose to your school. For example, you could choose a topic related to operating procedures, personnel, physical plant, or curriculum. Then write a letter proposal that persuasively puts forth your case for the change. As the proposal writer, place yourself in the role of an outside consultant writing to particular decision makers at that school.

2. Letter Proposal to a Company: Real Content

Write a letter proposal in which you, as a representative of a company, propose the purchase of a product or service by another firm. Select a product or service with which you are reasonably familiar on the basis of your work experience, research, or other interests. However, in the proposal do not use proprietary information from any real company.

Be sure the proposed service or product is modest enough to be described adequately in a letter proposal, as opposed to a formal sales proposal. Consider this proposal to be unsolicited; however, write it to a customer for whom you have recently done some major work. Because the previous project went so well, you're convinced that this proposal will be seriously considered if you're able to write convincingly about the customer's need and the benefits of what you're offering.

3. Memo Proposal to a Firm: Simulated Content

Read Exercise #4 in Chapter 1 and use the situation details in that exercise to write a memo proposal.

4. Memo Proposal to a School: Real Content

If you're a college or university student, select a school-related problem about which you can write a proposal. Your topic can relate to your school's operating procedures, personnel, physical plant, curriculum, or extracurricular programs. As the writer, use your actual role as a student. Your proposal audience will be members of the college community who would make a decision to accept your proposed solution. Although you may not actually submit the proposal to anyone other than your instructor, write it as if the decision makers addressed were going to read it. Discover as much as possible about these readers. If your audience includes a director of residence halls or an academic dean, for example, you may want to conduct interviews to get background information for your proposal.

5. Memo Proposal to a Company: Real Content

Write a memo proposal suggesting a change that would improve any aspect of any company with which you are now or have been employed. Based on your employment experience, for example, you could propose changes for the company's operating procedures, personnel, physical plant, products, or services. However, in the proposal do not include proprietary information of any real company.

As the writer, adopt the position you actually have or had in the company, or place yourself in another role if it would make the proposal more realistic. Consider as your audience the managers and experts who would actually make a decision to accept or reject your proposal.

9 Reviews

OBJECTIVES

- **Learn the goals of various proposal reviews**
- **Learn the factors for choosing and scheduling reviews**
- **Learn the rules for organizing and conducting a formal review**
- **Perform exercises using chapter guidelines**

Reviewing a proposal is crucial to its success. There are many opportunities to review a proposal, including the time before and after RFP release. Table 9–1 describes a variety of reviews that can be used to support proposal development. Of the indicated reviews, you should at least conduct a comprehensive formal review. The other listed reviews can be used to supplement the formal review.

Reviews are often described by colors. For example, the formal review is frequently called the red team review. (The "red" may or may not be a reflection of the red ink used to critique the proposal.) Other common color-coded reviews are the gold team reviews for storyboards and pink team reviews for early drafts of the proposal in advance of the red team review. However, there's no standard rule for the color coding. The use of a color helps define your reviews and gives your review teams a "personality," but it's certainly not a requirement for a winning proposal.

Beyond the obvious value that reviews can add to your proposal, reviews can provide the following benefits:

- ◆ They allow the involvement of talented technical and management people who may not have time to serve as proposal writers but have time to serve as reviewers.
- ◆ They can be used to break the proposal preparation schedule into a series of manageable tasks for the proposal team.

TABLE 9–1: Types of Proposal Reviews

Review	Activity
Strategy and baseline	Reviews proposal strategies and themes, and the technical, management, and cost baselines for best competitive advantage. (The strategies and baselines provide a starting point for developing detailed approaches to meet your customer's needs.)
Pre-RFP draft	Reviews a proposal draft written before the RFP is released. (Draft proposals can be based on similar RFPs released by the customer or others, or on other proposals in which you offered a similar product or service.)
Storyboard/mock-up	*Informal:* assists writer development of the proposal storyboards/mock-ups, and then evaluates them before the formal storyboard/mock-up review. (Assistance and review can be provided by proposal team management.) *Formal:* reviews each storyboard/mock-up for compliance with RFP requirements and applicable strategies and themes, and reviews graphics for effectiveness and creativity.
Post-RFP first draft (also known as a pink team review)	Reviews the first proposal draft to evaluate it before it's formally reviewed. (This review can allow the proposal team to fix proposal problems and prepare a better proposal draft before it's evaluated during the formal review.)
Formal (also known as a red team review)	Comprehensively reviews a mature proposal draft to verify its compliance with RFP requirements and proposal strategies, and to ensure it's accurate, competitive, credible, and persuasive.
Follow-up	Reviews the proposal draft that was revised after the formal review, and verifies that proposal deficiencies are fixed. (This review can include the entire proposal or focus on a portion of it.)
Cross-read	Reviews the proposal draft during the formal review to ensure that all facts and figures agree with each other through the proposal.
Cost	Reviews pricing section or volume to ensure that it's realistic and competitive and that it contains the substantiation required to support the bid price. (The proposed cost can be evaluated as part of the formal review.)
Graphic	Reviews to ensure that the graphics are creative and effective and meet proposal style guidelines. (This evaluation can review first draft graphics after the storyboard/mock-up stage or review all draft graphics during the formal review.)
Final sign-off	Reviews to ensure that all issues raised during previous reviews have been resolved and to approve the proposal (and bid price) for submittal to the customer. (This review is conducted by upper management and is normally held close to the proposal delivery deadline. Allow enough time in your proposal preparation schedule to make changes caused by this review.)

Scheduling and selecting a review are affected by the following factors:

- Available time and money—The more time you have to deliver the proposal, the more reviews you can conduct. The scheduling of the review can also be affected by the readiness of the proposal draft for review, the availability of people and money for the review, and the time and money available to revise the proposal after the review.
- Proposal complexity and length—Complex and long proposals can require large review teams, staffed with reviewers from many disciplines. These teams can involve many people, who take several days to review a proposal in a dedicated review area. In contrast, simpler and shorter proposals might be reviewed by one or several people working in a small conference room or at a regular office desk to review the proposal in less than a day.
- Proposal importance—If winning the contract is especially important, extra attention to the review might be appropriate, regardless of the proposal complexity or length. This emphasis can increase the number, scope, and formality of the reviews, as well as the size and composition of the review team. The value of the proposed contract isn't always the best way to judge the completion time or the importance of a proposal. A proposal for a $500,000 contract might require as much time and be as important to your company as one for a contract worth $5 million.
- Specific need—A review may be based on a specific need. For example, it can address a specific technical, management, and cost issue; or a specific proposal process, such as graphic or storyboard development.

> Reviewing and polishing the draft is the final step before production. As you review, challenge everything . . . . If it isn't quite right, change it. If it isn't necessary, get rid of it. Be your own toughest critic. Ask this question of every fact and feature in your proposal: *So what?*
>
> (Franklin Consulting Group, 1992–1996, p. 4-51)

RULES FOR HOLDING A FORMAL REVIEW

If you have time for only one review, conduct a formal one. The following rules provide guidelines to organize and conduct your formal (red team) reviews.

Review Rule #1: Select a review leader.

A formal review is more effective if it has an assigned leader, who keeps the review team on schedule, coordinates the documentation and collection of review comments, and leads a debrief of review results. Without a leader and a process to follow, the review can be disorganized and unproductive. The leader should have a solid understanding of the proposal process, a basic understanding of the proposed product or service, and the ability to translate the detailed review comments into summary observations and recommendations. Before the review begins, the proposal manager and review leader should meet to plan the review, addressing such issues as the team assignments, the review procedures and schedule, and the material to be provided to the reviewers. Review planning can be led by either of them, or the two can share the responsibility. To promote review objectivity, don't assign your proposal manager as the review leader. However, the proposal manager can be used to help the review leader run the review.

■ *Review Rule #2: Commit and assign reviewers.*

Before the review date, obtain the commitment of all reviewers to participate. Be sure that each proposal part has at least one assigned reviewer. For short proposals, each reviewer might review the entire proposal; for long proposals, a reviewer might be assigned to a specific proposal section or volume. If there was a prior review, use some of the same reviewers because they can provide review team continuity. Their involvement can help avoid the frustration of being told by a new review team that the guidance from a previous review was wrong and that you must make wholesale changes to your proposal. In contrast, if you use the same review team for a series of reviews on a proposal, you may miss fresh and effective ideas that can come from new reviewers.

■ *Review Rule #3: Prepare the review team.*

Before the review date, provide reviewers with materials that will allow them to prepare. These materials include the RFP, strategy plan, requirements matrix, and a brief written summary of reviewer assignments and how the review will be conducted and documented. Ensure that reviewers become familiar with this reference material—especially the RFP—before the review begins. During a review, it's more productive to have them *reviewing* the proposal than it is to have them *preparing* to review the proposal.

If a proposal draft is available before the review date, it can be helpful to have reviewers read it as they prepare for the review. However, the reality of a pressing proposal preparation schedule can prevent this from happening.

■ *Review Rule #4: Conduct a kickoff briefing at the review.*

Begin the review with a briefing that provides an overview of proposal issues and requirements, and details about the review process, assignments, and schedule.

In the proposal overview, describe the proposed product or service, major proposal strategies and themes, key customer needs that led to the proposal, and an assessment of the competition. Either the review leader or proposal manager could provide this information.

One could argue that the review team shouldn't be told what the strategies and themes are, because it's best to know if reviewers can discover them based on their own reading of the proposal. Regardless of your approach, the review debrief should address the adequacy of the proposal strategies and themes found in the proposal.

Have the review leader brief the following information to the team:

- Assignments for each reviewer and the review schedule
- Procedures and forms to be used for documenting and collecting review comments
- Instructions for making comments that lead to specific recommendations
- Instructions for ensuring proposal compliance with the RFP and strategy plan as cross-referenced by the requirements matrix

■ *Review Rule #5: Commit resources and time for the review.*

Both the proposal manager and review leader should ensure that the review team has all the resources it needs. This includes facility space for the review, copies of the draft proposal, reference materials and review forms, and other materials and supplies, such as computer hardware and software, pens, paper, food, and beverages.

The following are tips for preparing the draft proposal for review:

- Pagination and printing—Number all pages of the master draft copy and print the copies of the draft on three-hole-punched paper and one side only. Page numbers—even if handwritten—are useful in tracking review comments and distributing sections to the review and proposal teams. Three-hole-punched paper allows the review and proposal teams to use three-ring binders for easier access to proposal sections and better control of the draft. Plus, it's much easier to print the draft proposal on three-hole, prepunched paper than it is to print the draft on no-hole sheets and then punch the three holes by hand. By printing the draft on a single side only, you can find it easier to break the marked-up copies of proposals into separate sections for distribution to the applicable proposal writers.
- Layout—Avoid integrating text and graphics for the review draft unless you can do so without hurting the quality and completeness of the proposal content. Instead, insert the graphics as separate pages after they are referred to in the text. Merging text and graphics for the review can use time that is better spent developing proposal content. Plus, if the review leads to significant proposal changes that force a new layout, the text and graphic integration could prove to be a waste of time. However, this integration can make it easier to estimate the length of the proposal—an important issue for one that is page limited. With a little effort, however, you can still accurately predict proposal length without merging text and graphics.

If possible, conduct the review in a dedicated area, separate from the normal work space. This will reduce review team distractions and help the team concentrate on the review.

Expect a thorough review to take longer than you think it should, and give your reviewers a realistic estimate of the time they need to commit to the review. Schedule enough time to brief the review team, conduct and document the review, prepare and present a review debrief, and provide opportunities for direct discussion between the review and proposal teams. (Consider serving lunch to the review team in the review area. This arrangement can reduce the time needed for the lunch break and increase the time available for the review.) Some reviews may take only several hours, while others may need a 10-hour day or several days. As previously noted, a review schedule can be affected by many factors, including the length and complexity of the proposal and the number of reviewers.

In large proposals, guard against reviewer "fizzle out." This behavior is identified by declining reviewer thoroughness and concentration as the review progresses through the day. It can be caused by reviewer fatigue and the rush to meet an impractical review schedule. To avoid this problem, the review leader should monitor review progress and provide the reviewers a realistic assignment and schedule. Also, instead of asking all reviewers to evaluate the entire proposal, the review leader might assign reviewers to specific sections. This will help to avoid overloading reviewers with work and ensure that all proposal sections are reviewed.

Reviewers can also help the proposal team after the review. Raise this possibility when asking the reviewers to participate in the review, and keep this post-review support as a scheduling option.

■ *Review Rule #6: Review based on established evaluation criteria.*

> The red team review ensures that the proposal meets the following criteria:
>
> - All requirements are addressed
> - Points are clear and easy to evaluate
> - Text and graphics approach are consistent
> - The competition database is included
> - Winning strategies appear in appropriate paragraphs.
>
> (Jacobs, Menker, and Shinaman, 1990, p. 174)

Provide the review team with criteria to evaluate the proposal, and have reviewers judge the proposal through the eyes of the customer. At a minimum, direct the review team to conduct its review based on how well the proposal

- Complies with the evaluation criteria and proposal instructions specified in the customer's RFP
- Explains the who, what, when, where, how, and why of the proposed product or service
- Substantiates its claims and explains the benefits of the proposed product or service
- Reflects the winning strategies and themes based on your strategy plan

Although the review team can note grammar and spelling problems, it shouldn't do it at the expense of a substantive proposal review. Direct the team's focus on judging how well the proposal meets the preceding criteria.

Review Rule #7: Document review comments.

> Criticism should be positive and constructive. If the reviewers believe something is wrong, they should have a suggestion for making it right; if the reviewers don't have a suggestion for making it right, they don't belong there.
>
> (Helgeson, 1994, p. 195)

Although oral feedback from a review is important, the backbone of a formal review should be written comments that lead to specific recommendations. Based on the use of the review form in Figure 9–1, there are three major steps for processing review comments. (In Figure 9–1, the blocks of the form are labeled with the applicable step number.) The three steps include the following:

1. Document and distribute—Associate the comment with the applicable proposal section and RFP/strategy requirement. Use a logging system to identify comments by the reviewer and by the team (entire proposal, volume, or major section). If identified by major section or volume, the comments can be more easily grouped and distributed by proposal section for use by the applicable proposal writer. Also provide administrative information, including the reviewer's name and how the reviewer can be contacted, and the review date.
2. Explain and recommend—Clearly identify the problem and recommend a solution. It's not enough to note something is wrong; a reviewer needs to recommend a remedy.
3. Respond—After analyzing the comments, questions, and recommendations, the proposal team decides its response to the recommendations. You can expect that most recommendations will be valid; however, no matter how well intentioned or qualified, a review team can make mistakes by providing contradictory, misinformed, or simply bad advice. Therefore, judge each recommendation on its merits before implementing it. More details about the disposition of recommendations are given in Review Rule #9.

To supplement the comment forms, reviewers can make notes on the proposal draft copies they are reviewing. However, these notes should only be used to support the forms, not replace them. The notes can be used by the reviewers to organize their thoughts before completing the comment forms and to identify typos or other minor problems that don't deserve a form input. Have the review leader give the comment forms and the marked-up proposal copies to the proposal manager for distribution to the proposal team.

Review Rule #8: Conduct a summary debrief.

Have the review leader present a summary of the review findings. The debrief shouldn't be a detailed discussion of each comment form but, rather, a summary of key observations and recommendations. In the debrief, present the

<table>
<tr><td rowspan="2">Team Review Log #
#1</td><td colspan="4">Disposition #3</td></tr>
<tr><td>Accept:</td><td>Reject:</td><td>Optional:</td><td>Research:</td></tr>
<tr><td colspan="3">Proposal (Volume/Section/Pg. #):
#1</td><td colspan="2">Topic:
#1</td></tr>
<tr><td colspan="5">Applicable RFP/Strategy Plan Requirement (Section/Pg. #):
#1</td></tr>
<tr><td colspan="5">Comment/Question:

#2</td></tr>
<tr><td colspan="5">Recommendation:

#2</td></tr>
</table>

<table>
<tr><td>Reviewer Name:
#1</td><td>Reviewer Phone #/E-mail:
#1</td><td>Date:
#1</td><td>Reviewer Log #:
#1</td></tr>
</table>

FIGURE 9–1
Comment form for a formal review

following feedback for the entire proposal and major proposal segments (section/volume):

- Quantifiable score—Use a numerical (for example, 1 to 10) or descriptive (for example, poor, fair, good, very good, or excellent) score based on evaluation criteria from the RFP or your own creation. If possible, use an evaluation and grading system to simulate that of your customer. Learn as much as you can about how your customer will conduct the proposal analysis and pick a winner.
- Strength and weakness summary—List the major strengths and weaknesses, realizing that, no matter how weak a proposal draft is, it must have at least one positive feature. For each weakness, recommend a solution.
- Critical issues—Identify any issues that require immediate attention for the proposal to be competitive. For each critical issue, recommend a solution.

Have the review leader support the debrief with the use of visual aids, such as overhead transparencies. If possible, have the entire proposal team attend the debriefing; at least have key proposal team leaders attend. Caution the proposal team not to turn the review into a debate: let the review team first present its findings and then allow the proposal team to ask questions or make comments. If detailed discussion is required for a specific issue, do it after the main debrief is completed or arrange a separate meeting among the applicable proposal and review members.

Review Rule #9: Track the disposition of review recommendations.

> Once you have ruthlessly reviewed and revised your own writing, you will still need other reviewers to find problems and point out weaknesses you did not notice. Accept these review comments as an opportunity to improve your section. Other people look for and see things you don't, and they find things you can't. Think of review comments as the final opportunity to help your company win the contract and make yourself look a little better in the process.
>
> (Franklin Quest Co., 1995, pp. 7-15)

If time is spent reviewing a proposal, it's important to have a process that implements approved review recommendations. After receiving the comment forms, have the proposal team classify the disposition of each recommendation into one of the following categories (as they are labeled on the comment form in Figure 9–1):

- Accept—The recommendation is valid; the change is mandatory.
- Reject—The recommendation is rejected because it is wrong, inappropriate, or unfeasible.
- Optional—The recommendation is valid, but not critical; the decision to change is an option to be based on the preference of the applicable writer.
- Research—The recommendation could be valid based on further analysis or study; if valid, it is processed as a mandatory (accept) or optional change.

Compile a log of all comment forms and use this log to document the disposition of all recommendations. In the log, forms can be listed by the team or review logging number.

SUMMARY

There are many types of proposal reviews. One of the most useful is the formal review, which takes a comprehensive look at the entire proposal. Use the following rules to prepare and conduct a formal review, also known as a red team review:

1. Select a review leader.
2. Commit and assign reviewers.
3. Prepare the review team.
4. Conduct a kickoff briefing at the review.
5. Commit resources and time for the review.
6. Review based on established evaluation criteria.
7. Document review comments.
8. Conduct a summary debrief.
9. Track the disposition of review recommendations.

EXERCISES

1. Scheduling and Planning Reviews: Simulation

This exercise can serve as the basis for a group discussion, an individual oral report, or an individual written report. Considering the reviews listed in Table 9–1 and any others you think might be useful, identify and schedule the reviews you would recommend for the following types of proposal activity:

- An unsolicited 30-page, one-volume sales proposal with a two-week development schedule from proposal start to submittal.
- A solicited 100-page, one-volume sales proposal with a 30-day development schedule from RFP release to proposal submittal. There will be no pre-RFP proposal development work.
- A solicited, three-volume sales proposal (including a technical, management, and cost volume, each with 100 pages) with a 45-day pre-RFP preparation schedule followed by a 60-day proposal development schedule from formal RFP release to proposal submittal. During the first 15 days of the 45-day pre-RFP period, you will submit questions and comments to the customer about the draft RFP and by the end of the 45-day period you will complete a formal strategy plan, proposal outline, and proposal storyboards based on the draft RFP. During the 60-day period after formal RFP release, you will update your strategy, outline, and storyboard planning as required by the formal RFP and produce as many drafts of the proposal as you prefer.

For each scheduled review, describe the purpose of the review; the composition, size, and organization of the review teams; how the review is to be conducted, documented, and briefed; and the factors that influenced your choice, staffing, and scheduling of each review.

2. Reviewing a Formal Sales Proposal

With a group of four to six reviewers, conduct a formal review of the formal sales proposal example in Chapter 7. Organize your review team, review the proposal, document your comments on the forms based on Figure 9–1, and present the results of your review in an oral team presentation not to exceed 10 minutes. In the presentation summarize the following:

- Major proposal strengths
- Major proposal weaknesses
- Recommendations to resolve major proposal weaknesses

In addition, grade the proposal and provide the results in the oral presentation. (Grade the evaluation criteria using a 1 to 5 scale: 1 = poor, 2 = fair, 3 = good, 4 = very good, 5 = excellent.)

Evaluation Criteria #1: How Well was the "Story" Told?

Indicate the grade (1–5) for each of the following evaluation factors, and then develop an average grade for the six factors.

1. Who will do the work?
2. What work will be done?
3. When will the work be done?
4. Where will the work be done?
5. Why is the work necessary and why was the proposed approach offered?
6. How will the work be done?

Evaluation Criteria #2: Was the Proposal Readable, Credible, and Persuasive?

Indicate the grade for each of the following evaluation factors, and then develop an average grade for the six factors.

1. Was the writing clear, understandable, and logical; did it use proper grammar; and was there a consistent writing style?
2. Were the proposed approaches credible and effectively substantiated?
3. Were the proposed approaches clearly linked to specific benefits for the customer?
4. Was the proposal visually appealing (use of layout techniques, graphics, bulleted lists, etc.)?
5. Did the proposal use graphics effectively for simplifying and reinforcing ideas and stimulating interest?
6. Did the text effectively refer to and describe the graphics?

10 Editing

OBJECTIVES

- **Learn more about the editing process**
- **Learn to edit for style**
- **Learn to edit for grammar**
- **Learn to proofread for mechanical errors**
- **Perform exercises using chapter guidelines**

In the three-part editing process, adjust the proposal style, grammar, and mechanics to make your writing most persuasive. It can be a painstaking, tedious process that forces you to pick apart your prose, often when you'd rather just finish and deliver the proposal. However, you must persevere! Editing demands the same amount of attention as the other two steps of the writing process—planning and drafting.

In this chapter, we'll describe how to perform the stages of the editing process. However, the chapter is by no means a complete description of editing techniques and rules. Instead, it provides basic guidelines to be used with other reference sources and your experience.

THE EDITING PROCESS

> No one can write or dictate a first draft that is good enough to be a last draft. No one. Effective writers know that no matter how good they are, their first draft is filled with bugs. Furthermore, they know that the best way to write is to push out the first few drafts quickly, allowing ample time to edit and revise.
>
> (Weiss, 1990, p. ix)

Calling the edit function a process rightfully suggests that you should edit in stages. As summarized in Figure 10–1, there are three editing stages:

- In stage #1, refine the style. Style denotes the personal stamp you give your writing. You can adjust style by changing sentence structure, paragraph length, and word choice, for example.

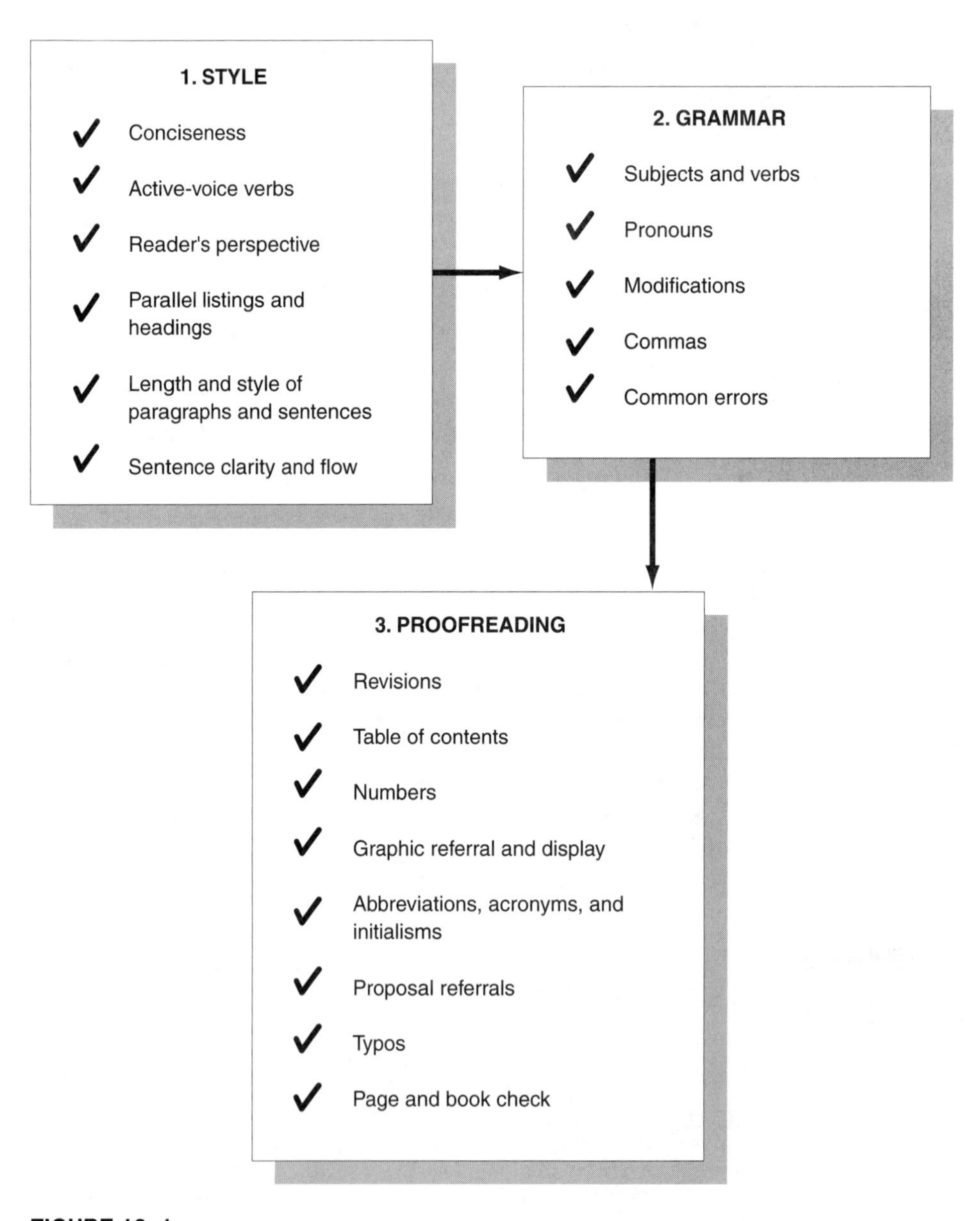

FIGURE 10–1
Stages of editing

- In stage #2, correct the grammar. Unlike style, which concerns matters of preference, this stage refers to matters of correctness. While editing for grammar, for example, you check to ensure correct usage of pronouns, punctuation, words, and agreement in number between subjects and verbs.
- In stage #3, proofread for mechanics. This step includes verifying that revisions are made correctly in draft text and graphics and checking to prevent other mechanical errors, such as misspellings, misplacement of decimal points, misnumbered pages, incorrect information in the table of contents, inconsistent use of numbers, and typographical errors.

LEVELS OF EDIT

> Polishing the draft means testing and fine tuning everything from the biggest, most blatant organizational problem to the comma you forgot to add; the word you didn't know how to spell; and the *to* that accidentally got an extra *o*.
>
> (Shipley Associates, 1995, pp. 7–17)

Breaking the editing process into three stages, or levels of edit, allows you to divide your editing into separate and manageable tasks rather than trying to complete all stages during one pass through the document. It also allows you to focus on specific types of errors. For example, if you often make subject-verb agreement errors, Stage #2 might involve one reading just for agreement and a second reading for all other grammatical errors. Or if you're a poor speller, you may want to back up your use of a word processing spell check function by reading the document several times just for spelling.

The levels of editing also allow you to tailor your edit when faced with schedule and budget constraints. For example, because there may not be enough time to do a thorough three-stage edit after the proposal has gone through a final review, you might choose only to edit for grammar and mechanics. This is the type of time-management decision that sometimes must be made to meet a proposal delivery deadline, although we don't recommend that you take shortcuts in the editing process.

SCHEDULING EDITS

Give your edits the attention they deserve by identifying them in your proposal preparation schedule. Schedule your edits in coordination with the proposal reviews. Focus the edit on readability, clarity, style, grammar, spelling, and format of the proposal. As was described in Chapter 9, use the reviews to assess how well the proposal implements your strategy plan and responds to the customer's RFP requirements. The review team can identify poor readability, grammar, or spelling, but it shouldn't be editing at the expense of its primary jobs. (Besides, reviewers are most often assigned based on their knowledge of the proposal topics, not their editing skills.)

Assuming you have limited resources and time for proposal edits, schedule edits when they'll be most effective. For example, you might delay having a full or even partial edit until the proposal draft has been reviewed at least once, because early drafts are often missing basic details or have major content flaws. It can be unproductive to edit a prereview draft thoroughly when the draft later undergoes extensive changes caused by the review. In another example, if you can conduct only one edit, it would be more effective to perform it after all reviews and the final draft are completed. You then can edit a mature version of the draft without your edit being superseded by a major content change caused by a review or author's late rewrite.

ASSIGNING AN EDITOR

Each proposal should be assigned an editor, whether that person is a professional editor or one of the proposal management leaders such as the proposal manager, volume leader, or proposal coordinator.

Even if you have a dedicated proposal editor, you can use other people to support this key task. In editing, too many cooks don't necessarily spoil the broth. The more sets of eyes that see a document, the better its chances of being edited well. However, to ensure editorial consistency in the final version of the proposal, you should have the editor-in-charge perform the last edit. Having one person edit a multiauthored proposal can lead to a consistent writing style throughout the proposal, making the document read as if it were written by one writer.

Others can be recruited to support your proposal editor. Volume leaders can be assigned to edit their respective volumes and then pass the volumes to the proposal editor. You could ask your boss—who's not involved in preparing the proposal—to read and suggest stylistic changes in a particularly important letter or part of a proposal, such as an executive summary. Or you could adopt a version of the "buddy system" by asking a colleague to review the draft for grammatical errors after you've done a thorough screening. Another version of the buddy system involves reading to someone else while that person follows along on another copy of the document. Although this approach takes time, it's a good way to catch errors.

Take full advantage of editing resources your firm offers. These resources can include proofreaders who can be used to find mechanical errors, although they may not be knowledgeable about the proposal topic or have strong skills for style editing.

Also consider hiring contract editors to support the three stages of editing. Using outside editors can provide the following benefits:

- Compared with your in-house employees, they may be able to assess more objectively the readability and clarity of proposal content because they aren't as knowledgeable about the proposal topics.

- If they edit the proposal so they understand it, it's likely that the proposal will be understood by a diverse customer audience.
- Employed only when you need them, they provide a capable source of editing support, whether or not you have in-house editing resources.

Editors, provided internally or externally, should know your style and format standards, the proposal outline and content requirements, and any other information that will help their edit.

EDITING TOOLS AND POLICY

To supplement an edit, the writer or editor should consider using the spell and grammar check features available with word processing software. At a minimum, require the writer to spell-check the draft before submitting it for edit. After editing, have the writer review the edited version to verify its accuracy. An editor can unwittingly change the meaning of a sentence by even minor editing. If questions were raised during the edit, have the editor and writer work together to resolve them.

Set style standards and editing policies to furnish consistent guidelines to the writer and editor and reduce conflict between the two in the editing process. A commercially—or internally—produced style guide can provide the writer and editor with the "final" word on style, grammar, and mechanics standards. The style guide can be supplemented with an approved list of selected words (with spelling and capitalization standards), abbreviations, acronyms, and initialisms.

> . . . you must accept critical editing. For some people, especially those for whom being "perfect" has an emotional charge and for whom criticism is an attack on self, this is difficult. But please understand that the best writers have to revise; it is almost impossible to write grant proposals well in one sitting. . . . You will benefit if you can free your defenses, not become rigid about redoing, and assume that feedback will be helpful.
>
> (Gilpatrick, 1989, p. 9)

Although it's hoped an editor and writer can resolve differences over an edit, there may be an impasse. Set a policy about who has the authority to accept or reject editing changes. For example, the editor might have final authority for changes based on objective grammar, spelling, and mechanics standards, while the writer has the final decision over style changes based on preference. If no agreement is reached, the decision can be made by the proposal manager. Editor-writer conflict can also be reduced if editing comments are written in a tactful, constructive manner. Avoid comments that only criticize; focus on recommending changes for improvement.

MARK	MEANING	EXAMPLE
^	insert	technical
ℓ	delete	technicaal
—	delete a word	at his his request
ℓ	delete and close space	ground water
/ or #	insert space	visit/England or visitEngland
∾	transpose	techincal
⁀‿	close up space	tech nical
≡	use capital	mcDuff, Inc.
=	use small caps	p.m.
/	use lower case	the Committee
V	add apostrophe or quotation marks or superscript	McDuffs policy
^	use comma here	McDuff Inc.
⊙	use period here	Inc
(;)	use semicolon here	St. Paul, Minnesota
(:)	use colon here	as follows
( )	use parentheses here	See Table 4.
[ ]	use brackets here	SIC
V=	use hyphen here	well planned meeting
¶	start new paragraph	. . . today. Then he began . . .
No ¶	take out paragraph change	He stated that the firm . . .
stet	keep original; disregard editing change	stet General Admiral Jones
]	move right	21 Walnut Street Portland, Ohio (216) 374-0011
[	move left	21 Walnut Street Portland, Ohio (216) 374-0011
⊔	lower	. . . the defense policy."3
⊓	raise	. . . the defense policy."3

FIGURE 10–2
Editing marks

MARK	MEANING	EXAMPLE
‖	align	.34 beams .12 bolts .15 bars .7 hammers
(Sp) or ○	spell out	(Sp) 3 team members or (3) team members
//////	remove underline	no significant pollution
1/M or M	add conventional dash	a big change 1/M and I mean big.
1/N or N	add small dash	1972 1/N 1982
⊃	run together	. . . gave his firm growth. Later he phased in . . .

FIGURE 10–2—*continued*

Determine if your editing will be done by keying directly in a word processing text file (on-line) or by handwriting (off-line) on a hard copy of the draft. With on-line editing, you can avoid the off-line steps of (1) printing and hand marking a hard copy of the proposal draft, (2) manually keying in changes based on the written editing marks, and (3) proofreading to confirm the changes were made correctly. You might prefer on-line editing if proposal text, graphics, and document files are being coordinated and transferred via an on-line networked system. The revision function in word processing software can also make on-line editing an attractive option. This function allows the editor to make editing changes directly into the file, with the changes displayed in color on the computer screen. With the file document displayed, these changes can be reviewed and then approved or rejected on line by the writer.

However, compared with viewing, scrolling, and editing an on-line document, the editor may find it easier and faster to read, comprehend, and mark up a printed copy of the proposal. If you edit off-line, use standard editing marks and symbols to avoid misinterpretation of the editing instructions. Figure 10–2 shows examples of commonly used editing marks. (By the way, writers can be less threatened and combative about editing marks and comments written in a color other than red.) An off-line edit should allow the writers to accept or reject editing changes before any changes are made in the computer files. This approach can help you avoid the input of editing changes into the file that must be deleted later because they weren't accepted by the writer.

Regardless of the number and scheduling of the edits and who performs the edits, there comes a time when the editing must end. Writers (and others on the proposal team) may want to "wordsmith" at the expense of meeting proposal production and printing deadlines. Therefore, set final and nonnegotiable limits for accepting changes to the proposal text and graphics. To relieve pressure on the final production process, use a staggered editing schedule. Instead of scheduling the edit of the entire proposal edit as one milestone, schedule edits for portions of the proposal—major sections or different volumes—as they're completed.

Remember that you're ultimately responsible for all material with your name on it, no matter who else may have helped its edit, so always make one last pass through any sales letter or proposal document before you sign it.

RULES FOR IMPROVING STYLE

> Your text can be quite accurate and thorough, although difficult to read. But prose can also be easy to read, while failing to be clear. For the client to be either puzzled by your meanings or misinterpret them is just as deadly to your purpose as discouraging the client's reading entirely. You can hardly sell something to the client when the client does not understand what you are selling or precisely what your arguments are.
>
> (Holtz, 1986, p. 177)

Like fingerprints, writing styles are unique. Throughout your life, you have developed your own writing style. Despite this individuality, a few basic rules of style apply to all good writing. The following rules for Stage #1 of the editing process are intended to make your proposals and sales letters conversational and persuasive.

■ *Style Rule #1: Be concise without sacrificing clarity.*

Long words, phrases, and sentences don't impress readers. Particularly in sales writing, you should write as you speak, and most of us speak with a simple and clear vocabulary.

Conciseness provides many benefits. It can make your writing more readable and understandable, and it can reduce the number of proposal pages—especially useful in a page-limited proposal. Reducing the proposal size can also reduce your production and printing costs. The following are ways to achieve conciseness:

- Use short words. Avoid a long, abstract word when a short, concrete one will do. In the following examples, notice how a long word can be replaced with a short one.

EXAMPLES

advantageous—helpful
alleviate—lesson, lighten
commence—start, begin
discontinue—end, stop
endeavor—try
finalize—end, complete
initiate—start, begin
principal—chief, main
prioritize—rank, rate
procure—buy, get
subsequently—later
terminate—end
utilize—use

- Shorten wordy phrases. Avoid long, wordy phrases, especially trite ones. The culprits are often long prepositional phrases that seem to do nothing but clutter sentences. Simplify your writing by finding shorter substitutes, as in the following before-and-after examples.

EXAMPLES

afford an opportunity to—permit
along the lines of—like
an additional—another
at a later date—later
by means of—by
come to an end—end
due to the fact that—because
during the course of—during
give consideration to—consider
in advance of—before
in the final analysis—finally
in the neighborhood of—about
in the proximity of—near

- Delete clichés. Clichés are expressions that help readers see the unknown in terms of the known—for example, costs that accelerate as quickly as a skyrocket ("skyrocketing costs"). Once overworked, however, a cliché can cease to inspire new images and just take up space. Clichés can also appear too informal in a proposal or sales letter and can be misunderstood by the reader. Avoid clichés such as those in the following examples.

EXAMPLES

a step in the right direction
as plain as day
ballpark figure
by leaps and bounds
efficient and effective
explore every avenue
few and far between
firm but fair
for all intents and purposes
heart of the matter
last but not least
leaves much to be desired
needless to say
reinvent the wheel
too numerous to mention

- Use clear subjects and action verbs. Avoid the use of "there are" and "it is," because these constructions delay the identification of who or what is doing something and can make your writing lifeless and abstract. Instead, use clear subjects and action verbs. Note in the following examples that the revised passages identify who's doing what in the subject and verb positions.

EXAMPLES

There are many Acme projects that could be considered for design awards.
Revised: Many Acme projects could be considered for design awards.

It is clear to the hiring committee that writing skills are an important criterion for every technical position.
Revised: The hiring committee believes that writing skills are an important criterion for every technical position.

There were 15 people who attended the meeting at the customer's office in Charlotte.
Revised: Fifteen people attended the meeting at the customer's office in Charlotte.

- Delete extra words and phrases. This guideline covers all wordiness errors not described earlier. Take out extra words and redundant phrasing that don't add a necessary transition between ideas or provide new information to the reader. (Remember, however, that it can be useful to repeat main points for emphasis.) The following examples show how revisions can trim sentences without detracting from their meaning:

EXAMPLES

Preparing the customer's final bill involves the checking of all invoices for the project.
Revised: Preparing the customer's bill involves checking all project invoices.

During the course of its field work, the team will be engaged in the process of reviewing all of the notes that have been accumulated in previous studies.
Revised: During its field work, the team will review all notes accumulated in previous studies.

Because of his position as head of the public relations group at Acme, he planned such that he would be able to attend the meeting.
Revised: As head of the Acme public relations group, he planned to attend the meeting.

The department must determine its aims and goals, so that they can be included in the annual strategic plan produced by Acme for the year.
Revised: The department must determine its goals to include them in the annual strategic plan.

■ *Style Rule #2: Use strong, active-voice verbs.*

English verbs are either linking, intransitive, or transitive. Linking verbs simply connect subjects with words that rename or modify the subjects, as in "George is an excellent writer." Intransitive verbs express complete action and thus are not followed by direct objects, as in "The printer failed." Transitive verbs transfer action from the subject to the direct object, as in "We will use a subcontractor's services."

The terms *active* and *passive* apply only to transitive verbs. When a verb is in the active voice, its subject performs the action. Somebody does something, as in "We will drill four borings at the site." The emphasis is clearly on the agent of the action. When a verb is in the passive voice, its subject receives the action of the verb, as in "Four borings will be drilled by us at the site." Here the emphasis is on the thing being done, rather than on the agent. Note that voice concerns the perspective from which the sentence is written, not the time (or tense) of the verb phrase.

From the examples in the two previous paragraphs, you can see how the following changes can occur when an active voice sentence is converted to the passive voice:

- The subject becomes the object of a prepositional phrase ("by us") or is dropped from the sentence altogether, possibly making it unclear who completed the action.
- The verb phrase is lengthened ("will be drilled" as opposed to "will drill"). The direct object in the active sentence becomes the subject in the passive sentence ("borings").

Passive voice sentences tend to be longer, less forceful, and often less clear as to agents of the action than do their active voice counterparts. Why, then, do writers use them so often? Putting the best face on it, some writers believe the passive voice sounds more "objective" and "scientific." Readers, on the other hand, can view passive voice writing as boring at best and confusing at worst, for it often fails to say who's doing what. Style Rule #2 responds to readers' needs by suggesting that most—not all—sentences be written in the active voice. Use active verbs in the following situations:

- To make clear who is responsible for an action ("Mark Swope, a member of our staff, will observe the installation.")
- To emphasize a company name—yours or a customer's ("Danzell Plastics needs three new extrusion machines for its plant in Atlanta.")
- To replace a top-heavy sentence with one that mentions a table or figure at the beginning ("Figure 1 includes an illustration of the main frame design.")
- To write as concisely as possible—active sentences are, by definition, shorter than passive sentences

Use passive verbs in the following situations:

- To break the monotony of active voice writing
- To reconcile that you don't know the agent of the action ("The present machines were installed more than 30 years ago.")

- To avoid the repetitive use of "I" and "we" in the text
- To show the action is clearly more important than the agent ("Billing statements for the project will be sent at one-month intervals throughout the five-year project.")

■ *Style Rule #3: Choose wording that reflects the reader's perspective.*

When writing the first draft or two, you naturally tend to present information from your own perspective. The proposal can contain many first-person pronouns and repetitions of your company name (for sales proposals) or department name (for in-house proposals). To change to the reader's perspective, replace first-person pronouns ("I," "we") with the second-person pronoun ("you") whenever possible. Also, subordinate the name of your own firm or department in favor of emphasizing the name of the reader's organization. The principle is simple: when readers see their own name or the name of their company, they're more likely to believe that you want to identify and meet their needs. Use the following example as a model for applying this style rule to your own writing.

EXAMPLE

Mongood Ambulance Conversions, Inc., is pleased to respond to your request for a proposal to provide three new van ambulances to the City of Nickelville, Texas. Mongood has been converting vans of all makes into full-service emergency vehicles for more than 10 years, and supplied your city with its first van in 1996. We are certain we can provide a product that will meet the needs of your medical units.

Revised: The City of Nickelville, Texas, needs three new van ambulances, as noted in your RFP of May 4, 2000. You also note that these full-service vehicles must be delivered by October 2, 2000, before the opening of the new emergency center at Gent Memorial Hospital. In 1996 you purchased your first van ambulance, a Mongood conversion that still serves the City. Now the new demands of Gent Memorial require even more sophisticated vans.

The first sentence of the unrevised example shows more concern for the writer than for the customer. The second and third sentences reinforce this tone by putting more stress on the features of Mongood than on the needs of Nickelville. The revision, on the other hand, starts out with the customer's name and the strong verb, "needs." It also acknowledges the very important reason for the October delivery deadline—the opening of the new emergency center. Even the reference to the previous purchase is now couched in language that shows the city's needs have changed, despite the good service of the first van. In short, the revised version adopts the reader's perspective.

■ *Style Rule #4: Create parallel form in lists and headings.*

Parallel form means simply that like items are phrased in a like manner. Parallel form allows readers to move quickly through your proposal, without being slowed by awkward shifts in structure. While editing for parallel form, pay particular attention to headings at the same level and to listings (both in sentences and in indented passages). The following example shows revision to ensure the parallel form of four subheadings.

EXAMPLE

Subheading A—Surveying the High School Principals
Subheading B—Evaluating the Survey Results
Subheading C—Petition the Board of Education
Subheading D—Recommend the Curriculum Action

Revised:
Subheading A—Surveying High School Principals
Subheading B—Evaluating Survey Results
Subheading C—Petitioning the Board of Education
Subheading D—Recommending Curriculum Action

The unrevised version contains an error in parallel form. To match the first two leadoff words ("Surveying" and "Evaluating"), the verbs "Petition" and "Recommend" need to be changed to noun forms ("Petitioning" and "Recommending"). The next example shows the revisions needed to have parallel form within a listing.

EXAMPLE

To assemble necessary information on the three computers, we plan to perform the following tasks:

- Write computer firms to request specifications
- Read related articles in recent periodicals
- Equipment demonstrations at local stores
- Interviews with local salespeople will be conducted

Revised:
To assemble necessary information on the three computers, we plan to perform the following tasks:

- Write computer firms to request specifications
- Read related articles in recent periodicals
- Request equipment demonstrations at local stores
- Interview local salespeople

The first two bulleted items on the list indicate that the writer wants to convey action by using such verbs as "Write" and "Read." However, the last two items switch to a noun phrase and then to a complete sentence. These errors in parallelism can be corrected by making the third and fourth items start with the verbs "Request" and "Interview," respectively.

■ *Style Rule #5: Vary length and style of paragraphs and sentences.*

Although there are no absolute rules for paragraph and sentence length, there are some general guidelines that can streamline your writing.

We have suggested earlier that both letters and proposals should use shorter paragraphs at beginnings and endings. In practice, you should avoid lengthy paragraphs in all parts of the document because a reader may skip them if they're too long. However, you should avoid the fragmented effect of too many short, staccato paragraphs. Here are some guidelines for editing paragraph style:

- Provide variety by using paragraphs of different lengths, usually within a range of 2 to 10 lines.
- Seek an average paragraph length of about 6 to 10 lines.
- Begin most paragraphs, particularly longer ones, with the main idea. Never bury an important point in the middle of a paragraph.
- Use short paragraphs (two to four lines) to emphasize main points or to provide important transitions, because short paragraphs tend to attract attention.
- Draw attention to groups of similar ideas by using lists signaled by numbers (when referring to steps or a sequence) or bullets.

Follow these guidelines as you edit for sentence style:

- Place the main point near the beginning of most sentences, avoiding long introductory phrases and clauses. Be particularly wary of long passive-voice sentences in which the main verb appears at the end.
- Focus on one main clause in most sentences (a main clause has a subject and verb and can stand by itself). When you string main clauses together with "and" and "but," you can dilute meaning and lose the reader.
- Put important ideas in short sentences.
- Achieve an average sentence length of about 15 to 18 words for proposals and 12 to 15 words for letters.

■ *Style Rule #6. Use sentences with clarity and a logical flow.*

To help reader comprehension, ensure that sentences have only one meaning and are organized for a logical flow of ideas. Although the writer knows what was originally intended, neither the writer nor the editor can be certain that an ambiguous sentence will be correctly interpreted by the reader. Therefore, during an edit, imagine yourself as a reader who doesn't have the knowledge—or access to the writer—to correctly resolve the ambiguity. If you can interpret a sentence more than one way, assume that the reader will choose the wrong one. Then rewrite the sentence until there can be only one conclusion—the correct one.

Logical sentence sequence, supported by lead-in summaries and transition statements, can help readers understand the rhetorical flow of your proposal. Use the "main principle of organization," more fully described in chapter 4, to present your major points in introductory sentences and paragraphs before amplifying with details. Be consistent in how you amplify ideas first presented in serial listings. For example, after listing four major benefits of your product, amplify those benefits in the same order they were first listed.

To guide the reader through major sections of your proposal, look for transition points in the text to follow the preacher's maxim: "First you tell 'em what you're gonna tell 'em; then you tell 'em; and then you tell 'em what you told 'em." This approach can prevent your readers from getting lost in your proposal, wondering where the document is leading them and why the content is important. The maxim is also useful for oral presentations.

RULES FOR CORRECTING GRAMMAR

> . . . errors of grammar, even if they only rarely distort your meaning, can be embarrassing. People who make errors of grammar are often judged as either lazy or stupid.
>
> (Weiss, 1990, p. 113)

Carefully edit to eliminate errors in grammar. Grammar, for our purposes, concerns the correct use of words and sentence structure.

The suggestions in this section involve matters of error, while most of the style rules covered in the previous section involve matters of preference. We'll describe five common types of grammatical errors; if you need a more thorough review of grammatical rules, consult a comprehensive grammar handbook.

■ *Grammar Rule #1: Make subjects agree with verbs.*

Verbs convey action or state of being. Subjects perform the action and, thus, answer the question of "who or what" performed the action. Plural subjects take plural verbs and singular subjects take singular verbs. Agreement errors occur when writers mix singular and plural subjects and verbs.

To correct subject-verb agreement errors, isolate every subject and verb combination in every sentence. This method allows you to move quickly through the document, focusing only on subjects and verbs. It also helps you determine the degree to which you have succeeded in choosing strong verbs to convey action in your proposal or letter. Use the following guidelines to make verbs agree with subjects:

- Use a plural verb when subject parts are connected by "and."

EXAMPLE

The transformer and switchbox are to be delivered by July 1.

(A compound subject, such as "transformer and switchbox," must take the plural form of the verb.)

- Make the verb agree with the subject nearest the verb when subject parts are connected by "or" or "nor."

EXAMPLES

Neither prices nor a product description is included in the brochure.

Neither a product description nor prices are included in the brochure.

- Make the subject agree with the verb even when the subjective complement is a different number. Sometimes called a predicate noun or adjective, a subjective complement renames or describes the subject and occurs in sentences with linking verbs such as "is," "are," "was," and "were."

EXAMPLE

The topic of the proposal is windows for the old dormitory.

(The subject of the sentence, "topic," is singular, whereas the complement, "windows," is plural. Thus the verb "is" must be singular to match "topic.")

- Use a singular verb even when a singular subject is followed by other nouns in phrases beginning with "as well as," "along with," or "in addition to."

EXAMPLE

The president of the firm as well as the procurement officer is to address the meeting.

(The sentence subject, "president," is singular and followed by a prepositional phrase. However, if you were to change "as well as" to "and," the resulting compound subject would require that you change the verb to "are," its plural form.)

- Usually use singular verbs with collective nouns. Your choice depends on the meaning of the collective noun. A collective noun, though singular in form, refers to a group of persons or things—for example, "audience," "committee," "crowd," "family," "team," and "public." When a collective noun obviously refers to a group as a whole, use a singular verb. However, when the collective noun refers to the members of the group as they act separately, use a plural verb.

EXAMPLE

The engineering team arrives by helicopter each morning.

The committee do not agree on where to place the rig. (If such a collective noun–plural verb combination sounds awkward, you could rephrase the sentence: "The committee members do not agree on where to place the rig.")

- Use a plural verb with irregular plurals, such as "data," "phenomena," "strata," and "analyses." A word that can cause disagreement among writers is "data." Although it's a plural noun, "data" is often used with a singular verb. (A traditionalist may find this usage wrong even if it has become acceptable by others.)

EXAMPLE

The data are to be evaluated by March 15.

(But if your meaning is singular and you want to avoid the singular form, "datum," rephrase the sentence: "A part of the data is to be evaluated by March 15." Here the subject is the singular "part," while " data" is only the object of the preposition "of." Thus, the verb must be singular.)

- Use a singular verb with indefinite pronouns—for example, "all," "any," "anybody," "anyone," "each," "either," "everyone," "nobody," "none," "one," "some," and "someone." The most prominent exception to this rule is that "all," "any," and "some" may be plural when the meaning is clearly plural.

EXAMPLES

Each of the designers is aware of the demands to be placed on the building.

All of the machinists have expressed a preference for the new schedule.

Grammar Rule #2: Check all pronouns to avoid unclear reference, agreement errors, and sexism.

Pronouns, such as "this," "it," "he," "she," and "they," are words that serve in place of nouns. As such, they are convenient for avoiding repetition, but they can also pose problems. The following techniques will help you use pronouns properly:

- Be certain that your writing makes absolutely clear what such pronouns as "this" and "that" refer to. Even better, rephrase sentences by turning these vague pronouns into adjectives.

EXAMPLE

Change This will be completed by next August ***to*** This phase of construction will be completed by next August.

- Check every pronoun for agreement in number with its antecedent (the noun to which it refers). Of special concern are the pronouns "it" and "they."

EXAMPLE

Change The company plans to complete their Wyoming plant by next April ***to*** The company plans to complete its Wyoming plant by next April.

- Never use male pronouns and adjectives, such as "he" and "his," to represent categories that could include both men and women. Overtly sexist language, though a convention for hundreds of years, gives some readers the impression that only men inhabit the category to which you refer. Speaking more practically, your sales writing may well be read by women professionals who might be offended by your use of masculine pronouns. Avoid this problem by using plural forms, shifting to a second-person pronoun ("you"), or by not using the personal pronouns altogether. (Using pronoun identifications of "he/she" or "he or she" is a remedy, but this approach can be distracting and wordy.)

EXAMPLES

For the plural form, change When a worker enters the site, he should wear his hard hat ***to*** When workers enter the site, they should wear their hard hats.

For the second person, change After selecting her insurance option in the benefit plan, each new nurse will submit her application to the Human Resources Department ***to*** Submit your application to the Human Resources Department after selecting your insurance option in the benefit plan.

For no personal pronoun, change During his first day on the job, any new employee must report to the company doctor for his employment physical ***to*** During the first day on the job, each new employee must report to the company doctor for a physical.

■ *Grammar Rule #3: Make all modification clear.*

The longer a sentence, the more likely it will contain modifiers—words or groups of words that moderate the meaning of other sentence parts. Incorrect use of modifiers can cause two main problems: dangling and misplaced modifiers.

When a modifying phrase "dangles," it usually begins or ends a sentence that contains no specific word for it to modify.

EXAMPLE

In designing the foundation, several alternatives will be discussed.

(Who is "designing" here?)

When a modifying phrase is misplaced, it seems to refer to a word that it obviously shouldn't modify.

EXAMPLE

Floating peacefully near the rig, we saw two humpback whales.

(The whales were "floating," not the observers, so the sentence must be revised.)

At best, dangling or misplaced modifiers produce a momentary misreading or a good laugh; at worst, they can leave the reader hopelessly confused about who's doing what in the sentence. Follow two strategies to avoid errors:

- Place all modifying phrases as close as possible to the word that they modify.

EXAMPLE

Leaving the office on Saturday, we should be able to arrive in Miami by Monday morning.

- Correct modification errors, found during the editing process, by reworking the sentence.

EXAMPLE

Change Using satellite surveying techniques and several ships, the position can be located by next Tuesday ***to*** Using satellite surveying techniques and several ships, we can locate the position by next Tuesday ***or to*** If we use satellite surveying techniques and several ships, the position can be located by next Tuesday.

■ *Grammar Rule #4: Check all commas.*

The comma is one of the most frequently used punctuation marks. The following are a few basic rules that cover the common usage of commas:

- Use a comma to separate words, clauses, or short phrases written in a series of three or more items.

EXAMPLE

The materials will be ordered from Gibson Supply, Wilson Photography, and Davis Wall Products.

(According to present usage, a comma should always precede the "and" in a series—the so-called serial comma.)

- Use a comma to separate main clauses joined by coordinate conjunctions, such as "and," "but," "or," "nor," "so," and "yet."

EXAMPLES

Many geologic faults may exist at the site, but we have not located any maps that display these faults. (The comma is needed because the conjunction "but" separates clauses, which are groups of words with both a subject and a verb.)

Many geologic faults may exist at the site but are not displayed on our maps. (In this sentence, there's only one clause and thus no need for a comma.)

- Set off nonessential modifiers with commas. A nonessential modifier doesn't greatly define or limit the word it modifies; it simply adds more information. To put it another way, dropping a nonessential modifier shouldn't greatly affect the central meaning of a sentence. Essential modifiers, on the other hand, limit the words they modify and aren't enclosed by commas.

EXAMPLES

The floodplain, which is located about five miles from the site, should not affect our construction plans. (nonessential modifier)

The floodplain that is five miles from the site should not affect our construction plans. (essential modifier) (The implication is that because there may be another floodplain, the modifier beginning with "that" is essential in pinpointing a specific floodplain.)

- Appositives expand the meaning of, or rename, the nouns that precede them. Nonessential appositives are enclosed with commas, while essential appositives are not.

EXAMPLES

Thomas Perch, author of a book on asbestos, will be one of the investigators. (nonessential appositive)

The word *asbestos* has become associated with major health hazards. (essential appositive)

- Use a comma to separate two or more coordinate adjectives that modify the same noun. Two tests help to determine if adjacent adjectives are coordinate: (1) you should be able to reverse them without a change in meaning and (2) you should be able to substitute "and" for the comma between them.

EXAMPLES

They found some black, thick liquid below the tank.

The appliance is a reliable, safe product.

- Use a comma after most introductory phrases or clauses that have five or more words.

EXAMPLE

Before applying the new salary structure to the other offices, the Human Resources Department will hold meetings with the employees.

- Follow conventional usage in placing commas in dates, geographical names, titles, and addresses.

EXAMPLE

November 3, 1995, was the date that Joseph Barnes, Jr., started the firm. Now Barnes, Inc., is a thriving business in eastern Oregon. (Note the need for commas after "1995" and "Inc.")

- Use commas in lists only if you wish to treat the entire list as a sentence. (Lists in the form of single words or short phrases need not be separated by punctuation.) There are several options for punctuating lists of phrases or clauses. Because lists are often used in proposals, we offer options for list punctuation in the following three examples.

EXAMPLE

The entire office staff should be able to

- Respond quickly to customer calls,
- Pass customer complaints to the field staff, and
- Document each complaint within a day.

(Here the nature of the series, with its three phrases, suggests that the list could be treated as a sentence. Regardless, depending on your firm's preference, this list could have been written without using the commas, the "and" before the last item, and the final period. As shown in this list, it's customary to capitalize the first word in each item.)

- In related punctuation use, semicolons can be used to separate items in a series when one or more of the items contain commas.

EXAMPLE

The entire office staff should be able to

- Respond quickly to customer calls, as well as to personal visits to the counter;
- Pass customer complaints to the line workers, supervisors, and diggers in the field; and
- Document each complaint within a day.

- The following items are sentences, some of which are fairly long. This suggests that each listed item be punctuated as a separate sentence, although you would also have the option of using no punctuation.

EXAMPLE

We were presented with three alternatives by the consultant:

- The building can be evacuated for three weeks while the sprinklers are installed.
- The workers can move from room to room while the sprinklers are installed section by section.
- The installation can be postponed.

- One commonsense guideline for commas supersedes all others: always use a comma when it will clarify meaning, whether or not a rule applies.

■ *Grammar Rule #5: Avoid common errors in word usage.*

A few words tend to be used wrongly by many writers. When contained in a proposal or sales letter, incorrect word usage can be confusing to readers and be construed by them as carelessness. It also can cause liability problems if errors in word choice commit you to doing something you can't perform. The following are words that should have your special attention:

Affect/Effect

"Affect" is usually a verb meaning "to influence." "Effect" is usually a noun meaning "result." "Effect" can also be used as a verb meaning "to bring about," but this usage often results in wordiness and should be avoided.

EXAMPLES

The hurricane did not affect his schedule.

The effects of the hurricane were significant.

The treasurer changed the company's check-cashing policy. (It's preferable to use "changed" instead of "effected a change in" for this example.)

All Together/Altogether

"All together" is used when items or people are being considered in a group or are working in concert. "Altogether" is a synonym for "utterly" or "completely."

EXAMPLES

The three firms are all together in their support of the agency's plan.

There were altogether too many car accidents at the corner for that intersection to be considered safe.

Alternately/Alternatively

As a derivative of "alternate," "alternately" is used for events or actions that occur "in turns." In contrast, "alternatively" is a derivative of "alternative" and is used when two or more choices are being considered.

EXAMPLES

While digging the trench, the technician will use a backhoe and a hand shovel alternately throughout the day.

We recommend the use of two operators. Alternatively, you could use one with a readily available replacement.

Anticipate/Expect

These two words aren't synonyms; their meanings are distinctly different. "Anticipate" is used to suggest or state that steps have been taken beforehand to prepare for a situation. "Expect" means only that you consider something likely to occur.

EXAMPLES

Anticipating that the contract will be negotiated successfully, we have hired three new engineers.

We expect to find four types of soil at the Georgia excavation.

Assure/Insure/Ensure

"Assure" means "to promise" and is used in reference to people. "Ensure" and "insure" can be synonyms that mean "make certain." However, the current preference is for "ensure," reserving the word "insure" for the context of insurance. "Assure" and "ensure" should be used with care in proposals, for they express a degree of absolute promise and certainty to which a writer or company could be held.

EXAMPLES

Be assured that the machinery will be delivered by June 1.
He will call to ensure that the delivery was made.
Our agent has agreed to insure us adequately for the job.

Between/Among

The distinction between these two words has become somewhat blurred in usage. We recommend the use of "between" when referring to only two items and "among" for three or more items.

EXAMPLES

The agreement was just between the two departments. No one at corporate headquarters knew about it.

The copy of the contract was distributed among all branch offices.

Complement/Compliment

As a verb, "complement" means "to add to, to make complete, or to reinforce." As a noun, it means "that which completes the whole." "Compliment" is a verb meaning "to flatter" or a noun meaning "flattery." As for adjective forms, "complimentary" means "free" or "related to flattery," whereas "complementary" means "that which adds to, makes complete, or reinforces."

EXAMPLES

The new landscaping will complement the building's facade.

The complement of six recruiters brought the department's staff up to its normal level.

The cash award to Jamie was the firm's compliment for his service.

The bidding firms were discouraged from sending complimentary gifts at Christmas.

Compose/Comprise

"Compose" means "to make up" or "to be included in." "Comprise" means "to include" or "to consist of." In other words, the parts compose the whole, whereas the whole comprises the parts. The phrase "is comprised of " is unacceptable usage and probably results from a corruption of the correct "is composed of."

EXAMPLES

The technical, management, and cost sections compose the proposal.

The proposal comprises three main sections on technical data, management information, and costs.

Continual(ly)/Continuous(ly)

These words have distinctly different meanings. "Continual" means "happening over and over, repeatedly," whereas "continuous" means "uninterrupted." Proper usage of these two words in your proposal can prevent you from inadvertently offering something that you never intended to provide.

EXAMPLES

Over a three-week period, they will continually visit the site to observe the progress of construction.

Once at the site, our engineer continuously observed the grading operation for two hours. (This usage suggests that the engineer watched the grading process for two full hours.)

Disinterested/Uninterested

"Disinterested" means "unbiased or impartial." "Uninterested" means simply "not interested." Given the great difference in meaning, don't confuse a reader by using "disinterested" when you mean "not interested."

EXAMPLES

We will seek a disinterested party to mediate the talks between the employee union and the management committee.

They were uninterested in responding to the RFP because of the potential liability problems.

Farther/Further

"Farther" refers to physical distance, whereas "further" refers to abstract or metaphorical distance.

EXAMPLES

The entire crew had to walk another three miles farther before locating the half-buried wreckage.

After he left, we worked even further into the night on the cost projections.

Hopefully

Use "hopefully" to mean "in a hopeful way" or "with hope." Don't use it as a substitute for "I hope" or "it is hoped that."

EXAMPLES

The proposal team waited hopefully for the customer's phone call.

I hope that the customer will call soon.

Imply/Infer

The easiest way to distinguish these words is to remember that the speaker or writer implies ("suggests"), whereas the listener or reader infers ("concludes"). This distinction also applies to the noun derivatives, "implication" and "inference."

EXAMPLES

He implied that the firm would probably give us the contract.

We inferred from his remarks that the contract would create some massive scheduling problems.

Principal/Principle

"Principal" as a noun means "head official" or "money on which interest is earned." As an adjective, it means "main, chief, major." "Principal" is often confused with "principle," a noun meaning "a basic truth or belief."

EXAMPLES

The firm assured the clients that one of its principals, Mr. James Fedderman, would be involved with the project.

They planned to pay off the principal on the loan.

The principal reason for bidding on that project is the experience it would provide.

He considered it a violation of his principles to work for firms that have been indicted.

RULES FOR PROOFREADING

This final editing stage, which is concerned with correctness in spelling, typographical form, and similar mechanical matters, is best done by people distanced from the proposal. We won't belabor the need for correct spelling, because we assume that you'll carefully scrutinize your proposals and sales letters for misspelled words. The following are important rules for proofreading.

> Proofreading is very demanding work. It requires great attention to details and a highly developed ability to spot problems. . . . Evaluators see mistakes as signs of ignorance or sloppiness.
>
> (Whalen, 1996, pp. 10–12)

■ *Proofreading Rule #1: Check the accuracy of text and graphics revisions.*

Always verify that text and graphics revisions were accurately made. This proofreading should compare the written change instruction to the revised version of the text or graphic. It should be performed by the person who made the change and then by another person as a backup.

■ *Proofreading Rule #2: Check for table of contents accuracy.*

Ensure that the table of contents has page numbers for sections and subsections corresponding to the actual text pages on which these proposal parts begin. Then, verify that it accurately contains the wording and sequence of the proposal headings and subheadings.

■ *Proofreading Rule #3: Check for number accuracy.*

Numbers are difficult to proofread and tend to be given too little attention. The result can be particularly unfortunate with regard to costs, because a costing error can be not only embarrassing to a proposing company, it could be financially painful. (A proposal can be considered a contract.) Check every single figure and computation for accuracy. Unless the customer says otherwise, avoid using the decimal points and the accompanying cents figures, because the additional detail invites more mistakes and can make the cost estimates look higher than they are.

■ *Proofreading Rule #4: Check for proper graphics referral and display.*

Make sure that the list of graphics (illustrations or tables) accurately contains page references and graphics captions and identification numbers. Second, verify that your document text contains accurate references to the figure and table numbers, and that graphics are displayed in the proper relationship to these references.

■ *Proofreading Rule #5: Check for the proper use and explanation of abbreviations, acronyms, and initialisms.*

An abbreviation is a short derivative of a word, whereas acronyms and initialisms consist of the first letter of each word of a multiword term or phrase. (An acronym spells out a word; an initialism doesn't.) Obviously, you need to define any abbreviations, acronyms, or initialisms used in your proposals or sales letters. However, if the shorter terms are used just a few times, consider writing them all out, especially if you think the reader may become confused by their use or fail to remember their meaning. For example, in a proposal with many pages between the use of an abbreviation, it may be best to use the complete term instead of that abbreviation.

Define derivative terms the first time they are used; spell out the long version and follow it with the associated abbreviation, acronym, or initialism in parentheses. Thereafter, the abbreviation, acronym, or initialism can be used without its definition. In a document that uses many of these derivative terms, define them in a proposal glossary.

■ *Proofreading Rule #6: Check for the proper form of numbers.*

There are no absolute, hard-and-fast guidelines for writing numbers. Your company, customer, or professional organization may have its own standards for expressing numbers as either figures ("92") or words ("ninety-two"). If not, the following are guidelines you can use:

- Generally, use figures for numbers of 10 or more and words for numbers less than 10, with exceptions as noted below.
- Don't begin a sentence with a number figure. Spell the number out or rephrase the sentence to avoid starting with a number. ("Twenty-five salespersons attended the workshop." "There were 25 salespersons at the workshop.")
- Use number figures to identify graphics. ("See Figure 4.")
- Use number figures when a sentence contains several closely related numbers, even though some may be less than 10. ("They hired 16 engineers, 5 secretaries, and 3 clerks.")
- When two numbers appear in succession in a phrase other than a series, write one as a word and the other as a figure. ("The crew consisted of 12 six-person teams.")
- Follow a number in words with its corresponding figure in parentheses only in legal documents. ("The hourly rate will be fifty-five dollars [$55].")
- In dollar figures, include cents figures only when exactness to the cent is necessary.
- Place commas in figures of four or more digits. ("The project will last 5,000 hours.")

■ *Proofreading Rule #7: Check for proper referral to other parts of the proposal.*

Confirm the accuracy of every text or graphic referral to a location or content in another part of the proposal. For example, if you write in Section 1 that Section 3.1.1 contains information about your cost accounting system, ensure that *it is* Section 3.1.1 and that *it does* describe the system. Errors can occur during proposal development when the organization and content of the proposal change and the referrals don't reflect the change. If you direct the reader to another part of the proposal, refer to the section, enclosure, attachment, or appendix identification, rather than to a specific page number. This approach will eliminate the need to verify the accuracy of page references if the proposal pagination changes during the edit and layout process.

■ *Proofreading Rule #8: Check for "typos."*

This check includes looking for the inadvertent mistakes that can plague a document. These errors can include a variety of "typos," including missing punctuation marks, letters, and words, as well as misspelled words.

■ *Proofreading Rule #9: Do a final page and book check after printing.*

After the proposal is printed, perform a page check for every proposal copy that will be sent to the customer. Confirm that the proposal copy correctly contains every page in the correct order and that there are no printing discrepancies, such as misaligned text or graphics, page smudges or washouts, and blank or wrinkled pages. If the pages are to be bound with a method that won't allow easy page replacement, complete the page check before the binding. For binding methods that allow easy page exchange, you can perform the page check after the proposal is bound.

Although a page check isn't the place for editing and proofing, be prepared to correct gross errors detected during this check. Have a contingency plan to correct and reprint proposal text and graphics even as you prepare to package the proposal for delivery.

After the proposal is bound perform a book check. Verify the integrity of the binder and ensure that the binding doesn't interfere with the reading of the document. If you use multiring binders for the proposal, verify that the rings close correctly. If they don't, the proposal pages can become separated from the binder during shipment.

You can also perform page and book checks for copies that won't be sent to the customer. To save time, however, do them after finishing page and book checks for all customer copies.

SUMMARY

Editing requires the same effort as the other writing steps. Careful editing requires that you attend to three stages: (1) refine the style, (2) correct the grammar, and (3) proofread for mechanics. These stages, also called levels of edit, allow you to focus on certain editing tasks because of special demands or to limit your edit because you don't have the time or resources to complete all three stages.

Edits should be identified in the proposal preparation schedule. The timing of your edits should be influenced by the maturity of the proposal draft and the schedule for conducting reviews. Although every proposal should have an assigned editor, the editing function can be performed by various people from internal or external sources. Having one editor can promote consistency of style in a multiauthor proposal. For consistent editing and reduced editor-writer conflict, set style standards and editing policies to follow.

Use the following rules to edit for style:

1. Be concise without sacrificing clarity.
2. Use strong, active voice verbs.
3. Choose wording that reflects the reader's perspective.
4. Create parallel form in lists and headings.
5. Vary length and style of paragraphs and sentences.
6. Use sentences with clarity and a logical flow.

Use the following rules to edit for grammar:

1. Make subjects agree with verbs.
2. Check all pronouns to avoid unclear reference, agreement errors, and sexism.

3. Make all modification clear.
4. Check all commas.
5. Avoid common errors in word usage.

Use the following rules to proofread for mechanical errors:

1. Check the accuracy of text and graphics revisions.
2. Check for table of contents accuracy.
3. Check for number accuracy.
4. Check for proper graphics referral and display.
5. Check for the proper use and explanation of abbreviations, acronyms, and initialisms.
6. Check for the proper form of numbers.
7. Check for proper referral to other parts of the proposal.
8. Check for "typos."
9. Do a final page and book check after printing.

EXERCISES

1. Sentences with Style Errors

Revise these sentences by following the style rules in this chapter. Some revisions may need to contain more than one sentence.

a. The conference last week in downtown Boston afforded us the opportunity to meet with several members of your staff in an effort to make plans to finalize the agreement between our respective firms.
b. Due to the fact that the request for proposal addresses the concerns that we have explored with your department chiefs on occasions too numerous to mention, it is our opinion that a meeting before the proposal is submitted might be the most efficient and effective way of exploring our mutual concerns and thereby discovering the various needs that you want covered in the document, and of arriving at a ballpark figure for the project cost figures.
c. It has become apparent that the above date is not a convenient time for the three firms to hold their joint discussions and it is therefore requested by our firm that another date be established so that the subcontracting discussions can be held in the very near future.
d. A series of step-by-step procedures, which have been established by our field technicians, are certain to give a great deal of assurance that the field work will be accomplished with the necessary accuracy, and the main ones are as follows:

   - All samples will be taken from the ground without being disturbed
   - Place the samples in plastic bags
   - Label bags
   - Bags containing the samples will be immediately sent to our Seattle lab for testing

e. In regard to your kind inquiry regarding the availability of our new pump in Florida, it is our pleasant task to inform you that the new pump is carried by two different dealerships, one in the city of Pensacola and the other dealer is in the city of Tallahassee.
f. We are pleased to submit our proposal for the Tri-City construction project that we hope to complete for your firm, Jones Engineering, next year and that we think will satisfy all needs we understood from the project description we recently received at our main office.
g. Conducted under the sponsorship of your firm and our own, the survey that was completed yielded results, as can be seen in Figure 8, that allowed us to prioritize the needs of the managers of all utility firms in the Chicago area.
h. A good deal of care will be utilized by us in determining the most advantageous design for the building referenced in your letter of July 21.
i. It was noted by this writer that several vats of pellets, as a viable alternative, could be shifted to the other production line, as can be seen in the diagram contained in Figure 4.
j. In the neighborhood of about 400 hours were spent by our road crews in an effort to repair the stretch of Route 45, but we are not yet in the position to give our exact determination of the manner in which this amount of time can be effectively reduced on the next repair project.

2. Sentences with Grammar Errors

Revise these sentences by following the grammar rules in this chapter. Be able to explain the rationale for each change you make.

a. His main concern is that the graph, when completed will include all data that was collected at the site.
b. The group from the three mid-town companies are planning another meeting.
c. Cabot Engineering, Inc. announced in their company newsletter that each department supervisor must submit his annual report by next Tuesday.
d. Barry Shockley author of the study, along with his colleague Kevin Black are planning to contribute to the project.
e. Because they could not agree on the purchase the three partners decided to seek outside advice.
f. I submitted a thorough well-edited report before the deadline and I was convinced my boss would like it.
g. The plan unless we have completely misjudged it will increase sales markedly.
h. The new personnel database will help us to:
   - Control costs of healthcare;
   - Monitor the training our employees receive;
   - Complete Affirmative Action reports.

i. Each of the proposal coordinators selected their own team for the in-house competition.
j. Either the staff accountants or the office manager are planning to review the cost proposal.

3. *Sentences Concerning Word Use*

Select the correct word in parentheses and be able to defend your choice.

a. The decision the president makes next week will be (affected/effected) by the many conversations he has with his managers this week.
b. It was (all together/altogether) too late to make a change to the cost proposal.
c. The first option is the total demolition of the building. (Alternately/Alternatively), the second option is the major renovation and repair of the building.
d. The company (anticipates/expects) that 10 new technicians will be hired this year.
e. He wants to (assure/ensure) his boss that the proposal responds to the customer's needs.
f. To (ensure/insure) that our liability will be limited, we (insured/ensured) the machine for $10,000.
g. Before making the decision, the project manager discussed the issue (between/among) the three department managers.
h. The advertising agency regularly sends out (complementary/complimentary) gifts to all its clients.
i. The company library (is comprised of/comprises) over 5,000 books. Surprisingly, almost 4,000 of them (compose/comprise) the country's largest collection on geological fault zones.
j. She (continually/continuously) worked on the tables and figures for several days, stopping only for meetings with the drafters and computer graphics specialists.
k. After working four years on just one building design, she became (disinterested/uninterested) in the project.
l. He planned to read (farther/further) after taking a break for lunch.
m. (I hope/Hopefully,) this proposal will be a winner.
n. John had tried to (imply/infer) in his speech that salary raises would be low, but his staff did not receive that (inference/implication).
o. His (principal/principle) concern was that the loan's interest and (principle/principal) remain under $50,000.

11 After Proposal Submittal

OBJECTIVES

- **Learn to respond to customer questions after proposal submittal**
- **Learn to prepare and deliver oral presentations**
- **Learn to prepare and use presentation graphics**
- **Learn to minimize nervousness during oral presentations**
- **Learn winning negotiation techniques**
- **Learn strategies for best and final offers**
- **Learn to perform post-award activities**
- **Perform exercises using chapter guidelines**

The proposal process doesn't end when the customer receives the proposal document. Your work *after* proposal submittal can be the deciding factor in winning with a well-written and persuasive proposal—and even with a not-so-good proposal.

This chapter describes tasks that can occur after proposal delivery: (1) responding to customer questions about the proposal, (2) giving an oral presentation in support of the proposal, (3) negotiating a contract based on your proposal, and (4) performing actions after an award decision. The relationship of this post-submittal activity is summarized in Figure 11–1.

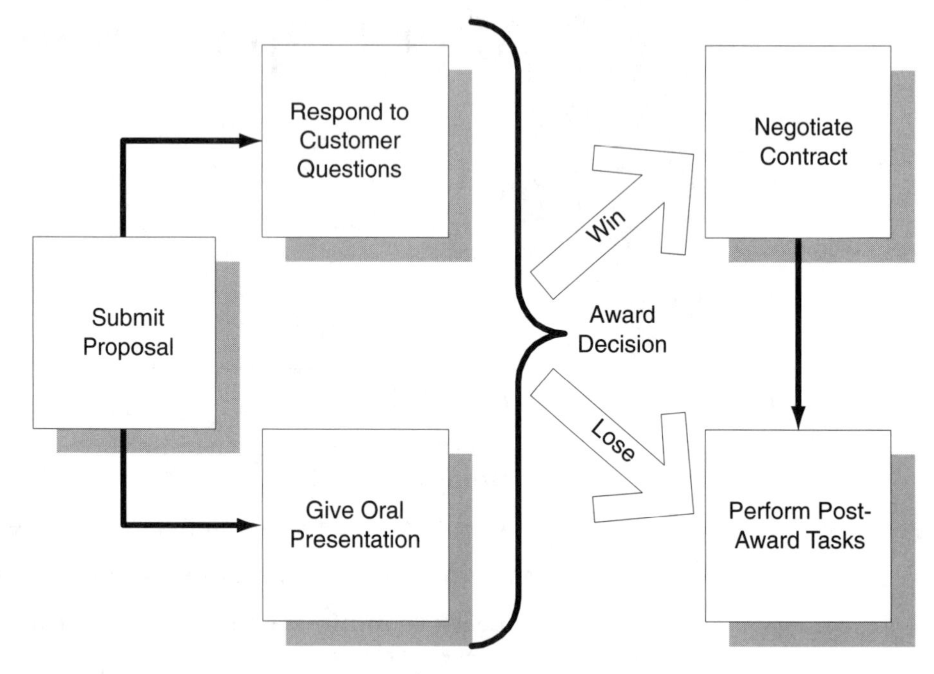

FIGURE 11–1
Post-submittal activities

RULES FOR RESPONDING TO CUSTOMER QUESTIONS

During the evaluation of your proposal, the customer may ask you to explain something not covered in your proposal or to clarify something that is. Use the following rules to respond to these questions as diligently as you developed the subject proposal.

■ *Response Rule #1: Have an established contact point for the customer.*

You must be able to receive and respond quickly to customer inquiries after proposal delivery. Don't lose a proposal competition because you didn't answer customer questions that were "lost" in the mail or fax room or in someone's basket.

To avoid a communication gap, designate a specific person to serve as the primary contact point for the customer. Unless otherwise directed by the customer, identify and explain how to contact this individual in the transmittal letter or main body of the proposal. Using a primary contact streamlines the communication process and gives your company the appearance of "one voice." With this approach you can quickly react to customer inquiries and also avoid conflicting statements that can occur when more than one person is speaking for your organization. Also, designate someone to serve as a substitute for the primary contact. Ensure that both are aware of their assignments.

■ *Response Rule #2: Assign responsibility and a deadline for the response.*

In addition to assigning a customer contact, designate someone to be responsible for answering customer questions. This person may or may not be the primary contact. Expect the questions to come with a customer-imposed deadline for your response. Have your response leader coordinate the development of your answers, ensuring that they are delivered to meet the submittal deadline—whether it's set by you or the customer.

■ *Response Rule #3: Answer completely and tactfully.*

> The winning proposal *takes advantage of questions and challenges to strengthen its position.*
>
> (Edmunds, 1993, p. 11)

To answer the question completely, you must first completely read the question. If you're not sure what it means, either preface your answer with your interpretation of the question or ask the customer to clarify it before you send an answer.

Look for explicit and implicit meanings in the questions:

- For the explicit question—Answer only what is asked; avoid tangential and unneeded responses that could contradict other proposal information or raise new questions.
- For the implicit question—Decide if the customer has worded the question or comments to rethink your approach or to suggest the "right" approach. For example, the question may suggest that a proposed technical design isn't appropriate and might even offer hints about a preferred design. Or the customer may submit a question that leads you to address a key issue or RFP requirement that you failed to address in the proposal.

Avoid answers that appear condescending or argumentative; don't let your response imply that the customer did a poor job of understanding or finding the answer in your proposal. Although it can be frustrating to know that a question was well covered in the proposal, don't let your frustration come across as sarcasm or an insult. Instead, tactfully note where the answer is located in the proposal and then restate the answer for clarity. If it wasn't covered in the proposal, don't apologize or make excuses for the missing information; just answer the question accurately and completely.

■ *Response Rule #4: Establish a review and approval process for all answers.*

Review all responses for accuracy, consistency, completeness, grammatical correctness, and tactfulness. Sometimes a question can be interpreted more than one way; a review can help you catch answers that fail to answer the real question. Schedule enough time to conduct a thorough review and to rework answers if required. Submit your answers only after they've been thoroughly reviewed, edited, and approved by your organization.

Be careful that your responses don't change the validity of your cost proposal—unless a cost change is authorized by the customer and you truly want to propose a new cost. To look for potential impact on your cost offer, note any response that proposes a change in such things as the schedule, the work scope, or the design or materials for your product.

■ *Response Rule #5: Provide answers to meet customer submittal instructions.*

After approving the answers, package and deliver them according to the customer's instructions. A common response method is to simply provide answers for the list of questions. If it doesn't conflict with customer instructions, submit your written answers along with their associated question. This approach will allow the reader to understand your answers without having to refer to a separate copy of the questions.

The customer may require changes to your proposal document to reflect your answers. If so, ensure that you carefully plan this work to meet the submittal deadline. As required, use special markings or notations in your revised proposal to identify the changes. A common method is to use a vertical bar in the page margin next to the change. Keep the proposal within its page allowance if the customer still maintains a proposal page limit that includes pages added or changed because of your responses.

Comply with all customer instructions for the delivery of your response, including the answers and any required proposal changes. As you should have done for the original proposal, have a backup delivery plan.

ORAL PRESENTATIONS

> Your purpose is determined by what you want your audience to do after you have made your presentation. Do you want your audience to have more information so that they will be more knowledgeable about your subject, or do you want to persuade your audience to act on, accept, or approve the information you present? The two main purposes of oral presentations delivered as part of the proposal process are to inform and to persuade.
>
> (Bowman and Branchaw, 1992, p. 174)

Customers often require an oral presentation to support the written proposal after it's delivered. Before you begin to develop this oral presentation, learn the conditions and requirements that will shape the presentation. Table 11–1 provides questions that can help you plan your oral presentations.

TABLE 11–1: Questions for Planning Oral Presentations

Topic	Basic Question	Specific Questions
Format and media	What is the format and media support approach?	• Will it be an informal question-and-answer session or a formal speech with or without customer questions? • What presentation media will be used—overhead transparencies, videotape, computer-based slides, flip charts, or props?
Scheduled start time	When is the scheduled day and start time?	• When will the presentation start? • Is the time dictated by the customer, or do you get to pick the time? • Will your competition also be speaking, and where's your place in the presentation schedule?
Duration	What is the length of the presentation?	• Is there a minimum and maximum time range for the presentation, or will there be only a maximum allowable period? • How much time will be dedicated to your presentation and to a question-and-answer session?
Presenter(s)	Who is to make the presentation?	• Will it be a solo performance by the proposal manager or a group presentation by the proposed project team? • Will your presenter(s) need assistance from those who won't be part of the prepared presentation but must be available to answer questions?
Content	What is the content of the presentation?	• Will it provide brief highlights or an intensive examination of the written proposal? • Will it support a physical demonstration of your product or service or a customer visit at your facility?

continued

TABLE 11–1: *continued*

Topic	Basic Question	Specific Questions
Stage in the evaluation process	When is the presentation to be made in the evaluation process?	• Will it be presented shortly after proposal submittal before any competitor rankings have been made by the customer? • Will presentations be made only by competitors who have survived a customer down-select process?
Location facility	Where is the presentation to be held?	• Will it be held at your facility or at the customer's facility? • When will the facility be accessible to you to preview and to set up for the presentation?
Presentation resources	What presentation resources will be available for your use?	• Will you have to provide your own presentation equipment, such as projectors, screens, and flip chart easels? • If the equipment is provided by the customer, when and how will it be available to you?
Support materials	What supporting materials will be required for the presentation?	• Will a paper copy of the presentation media, such as transparencies or slides, be required for distribution to the audience? • Will a file copy of the presentation media be required on disk or CD-ROM?
Audience	Who will attend your presentation, and how much will the audience know about your proposal?	• How familiar will the audience be with the content of your proposal? • Will the audience be the decision makers for awarding the contract? • What is the expertise and interest of your audience?

The RFP is a common source of oral presentation requirements. If you're unclear about customers' requirements, based on the RFP or any other source, ask them for clarification. Rather than setting stringent requirements, the customer may give you leeway in choosing your presentation approach. For example, you may be allowed to pick the date and time of the presentation, design the content of the presentation, choose the media of the presentation graphics, and select the speakers. Of course, if you're able to make these types of decisions, choose the approaches that best support your proposal and the people who'll make the presentation.

When you know the presentation guidelines, how do you develop and deliver the presentation? The following rules are for both the novice who needs a starting point and the experienced speaker who needs only a refresher. They focus on oral presentations to customers after proposal submittal. However, many of the rules can apply to all oral presentations, including those given to your own proposal team during proposal development and to the customer before proposal submittal.

We first describe rules for preparing and delivering the oral presentation, followed by rules for developing and using graphics. Finally, we offer tips for reducing the nervousness that can accompany your presentation.

Preparation and Delivery Rules

Start your presentation planning as early as possible. Just because the presentation will be made after you submit the proposal doesn't mean that you should start preparing for the presentation after proposal delivery.

In preparing a speech, remember that listeners

- Probably don't have your knowledge of your product, service, or idea.
- May have heard so many bad oral presentations that their expectations for yours may be low.
- May tend to daydream during your oral presentation.

Plus, there's a good chance that they won't be able to "rewind the tape" to review what you've said.

Given these obstacles, prepare and deliver your oral presentations with the following rules.

■ *Oral Presentation Rule #1: Use the Preacher's Maxim.*

As previously noted in this book, this maxim states, "First you tell 'em what you're gonna tell 'em, then you tell 'em, and then you tell 'em what you told 'em." Why do good speakers follow this plan? Because it gives the speech structure. Obvious beginning, middle, and end sections help keep the listeners oriented within the presentation. Here's how the maxim should work in practice:

- Beginning—In the first 30 seconds, state the precise purpose of the presentation—no matter how obvious it seems to you—and the main points you will cover. For example, you could say "Today we'd like to stress three main benefits you'll derive from a new field safety plan: lower insurance premiums, less lost time from accidents, and fewer employee lawsuits for unsafe job sites."
- Middle—Then methodically cover your main points in the body of the speech, making certain to provide obvious transitions ("Besides lower insurance premiums, a second benefit of the new safety plan will be higher morale.")
- End—Finally, summarize the talk by again referring to each of the main ideas ("In conclusion, you can benefit from this new plan because. . . .")

This approach responds to the needs and expectations shared by many listeners, whatever their profession or level of education:

- The strong start grabs their attention and gives them a road map for understanding the rest of the speech.
- They receive reminders to keep their attention, both because they can't move back and forth within the speech, as they can within a written document, and because of their natural tendency to lose concentration.
- A strong finish wraps up the speech and draws attention to points the speaker considers most important. People remember first what they heard last.

The selective repetition ensures that you won't leave your readers behind. Don't worry about the better listeners being put off by repetition, because many listeners appreciate confirmation of the main points they think they heard, and repetition can give them that assurance.

■ *Oral Presentation Rule #2: Plan your presentation to stay within the allotted time.*

Plan your presentation with a target length in mind, using graphics as the backbone of your presentation. Similar to the planning you would do for the written proposal, use storyboard techniques to conceptualize the presentation graphics.

A useful rule of thumb is to make graphics that will be displayed for an average of about two minutes per graphic. This timing can help you avoid rushing too quickly through graphics or getting stuck on a graphic because it requires a long explanation. This two-minute rule can also help you avoid using projected visuals that are too detailed for the audience to read. It can also lead you to produce only those visuals that you can comfortably use within the time limit of your presentation. For example, to prepare a 15-minute presentation, you could expect to use about seven or eight visuals. In this case, it would be a waste of time to produce 40 slides if you don't have adequate time to show them. The next section of this chapter provides more details about developing and using presentation graphics.

■ *Oral Presentation Rule #3: Use notes.*

Always bring presentation notes with you, even if you've memorized your presentation. They can keep you on track and cue you along if you get derailed by interruptions or just simply forget what you want to say. Also, occasionally glancing down at notes looks more natural than speaking as if the presentation were memorized. There are several ways to prepare your notes; choose the approach that best fits your presentation style.

You may want your notes on small note cards or on a single sheet of paper. Note cards can be easily carried in a coat or shirt pocket and can be held in one hand if you move away from the lectern while speaking. Compared with note cards, a single sheet of paper with notes or an outline can give you less material to handle and allow you at a glance to see the entire speech structure on one page. With less handling of notes, your hands can be freed to make gestures that will help your delivery.

Instead of written notes on paper or cards, you may prefer to use your visuals as notes and cues. If you're projecting transparencies or computer graphics, glance down at the overhead projector or the computer screen to look at the graphic. This may appear more natural than looking at the graphic projected on the audience's screen. Plus, if you look at the audience screen, you run the risk of turning your back to your audience. You can also make hard paper copies of each graphic and annotate them with written notes for referral during the presentation. Your presentation software may give you the option to print the graphics in miniature, allowing you to display several graphics on the same page. These printouts can also be annotated with written notes.

■ *Oral Presentation Rule #4: Practice, practice, practice.*

After you prepare the presentation content and graphics, follow through with an essential ingredient: practice. Frequent practice can separate superior presentations from mediocre ones. It will help you minimize the nervousness that speakers often feel before and during the presentation and will help keep you within your allotted presentation time. If you can't stay within your time limit during practice, expect the same problem during the real presentation. In addition to helping individual performance, practice can improve the transition from one speaker to another in a team presentation.

The following are ways to practice your oral presentation:

- Videotape the presentation. With this method, you can see yourself truly as others see you. At first it can be an unsettling experience, but with effort you can quickly get over the awkwardness of seeing yourself on tape. Careful review of the tape, followed by retaping, can help you improve your posture, gestures, vocal patterns, eye contact, and body movement.
- Record the audio of the presentation. This technique provides an inexpensive way of recording the presentation audio if you don't have access to a video recorder. It particularly helps you discover and eliminate verbal distractions, such as filler words.
- Practice before a live audience. It's helpful to practice before groups of your business colleagues, simulating the actual customer audience, or before friends or family members. Have the audience ask questions, perhaps even provoking confrontation to test your knowledge, tact, patience, and poise under stress.

Critique the practice presentations. Evaluate how well the presentation meets the customer's requirements. Also assess how well the presenters deliver the presentation and transition through presentation graphics and shifts in speakers. Treat this evaluation as you would for a review of the written proposal. Consider using a professional speech coach to critique and improve delivery skills of the presenters.

■ *Oral Presentation Rule #5: Speak vigorously and deliberately.*

All listeners expect to be kept interested. Good, relevant information helps, but the enthusiasm with which you deliver it is equally important. Listeners tend to give the benefit of the doubt to presenters who show energy. How much enthusiasm is enough? Here is a guideline you can use: speak with just enough vigor that the speech sounds a bit unnatural to you—that is, more enthusiastic than you would be in a one-on-one conversation on the same topic. That extra bit of zip may be just what the audience needs to sustain its interest. After all, you're giving a type of performance.

Your enthusiasm can help reduce the natural lethargy listeners develop in their passive role as an audience. It doesn't mean talking fast, however. Listeners can be suspicious of fast talkers, so speak deliberately, slowing down for the main points you want to emphasize. Another feature of deliberate delivery is the proper use of pauses. Well-timed pauses in your speech and physical movement help set up follow-on statements, add suspense, and give the audience time to digest a statement. Pauses are especially effective right before a major point, such as a main benefit, and can signal a change from one main point to another. The breaks caused by changing the display of your graphics give you other opportunities to pause.

■ *Oral Presentation Rule #6: Avoid those awful fillers.*

Filler words and phrases in presentations are like spelling errors in written proposals: once your audience finds a few, they may start looking for more. As a result, your message may have to compete for audience attention. The following are some of the most annoying fillers:

- "uhhh" (the most common culprit)
- "ya know" (used more often in the past few decades)
- "okay" (usually abused as a transition)
- "uhh . . . mm" (another version of "uhhh")
- "well" (a deep subject, as the cliché goes)

How can you rid your speech of these distractions? One technique is to record your practice, as mentioned in Oral Presentation Rule #4. When you play back the audio or video recording of the presentation, you can quickly note the fillers that occur frequently. The recordings will be brutally honest. Another approach is to practice before a live audience and have the listeners interrupt you when they hear fillers. This technique gives immediate feedback.

■ *Oral Presentation Rule #7: Use rhetorical questions.*

Next to enthusiasm, what vocal technique best captures listeners' attention? It's probably the rhetorical question.

Rhetorical questions, such as the one in the preceding paragraph, are questions you don't want the audience to answer out loud. Instead, these questions prod listeners to think about your point. They also prepare the audience for an answer you'll provide, as in this example: "You're probably asking, how much will it cost to install this new heating and air-conditioning system throughout the plant?" Then you answer your own question by emphasizing the cost-effectiveness of your product.

Rhetorical questions work well because they can break the monotony of standard declarative sentence patterns and establish listeners' expectations that an important piece of information is coming. Use rhetorical questions when you want to redirect or regain audience attention. These questions are most effective at the beginning of the talk, at transitions between main points, and immediately before the conclusion.

Oral Presentation Rule #8: Maintain frequent eye contact.

> Even when listeners are not saying anything verbally, they will be saying plenty with their bodies, eyes, faces, and tones of voice. Failure to attend and respond to the listeners' nonverbal messages can lead to their communicating through another nonverbal channel—their feet, as they walk out of the room.
>
> (Leech, 1993, p. 255)

Frequent eye contact helps you control the audience. Listeners tend to pay closer attention when you look at them. Here are ways to make eye contact a natural part of your delivery:

- With small audiences (fewer than 30 people)—Make regular visual contact with every person. Use a scan pattern that has you looking at all parts of the room.
- With larger audiences—Focus on four or five friendly faces in different parts of the room, giving the appearance that you're surveying the entire audience.
- With any size audience—Occasionally look away from the audience to collect your thoughts and to avoid the monotony of intense staring. Look either toward your notes, as mentioned in Oral Presentation Rule #3, or toward a part of the room where there are no faces.

Oral Presentation Rule #9: Use appropriate gestures and posture.

Integrate appropriate gestures into your delivery approach, being careful to avoid those that can annoy and distract your audience. Table 11–2 identifies both bad and good gestures.

Oral Presentation Rule #10: Use humor carefully.

Use mature judgment when using humor in the presentation. Humor can be an effective way of connecting with the audience and for illustrating points. However, it can appear as a forced and sophomoric attempt at comedy. It can also offend your audience; it might be funny to some but insensitive or insulting to others. Your knowledge of the audience and a strong sense of tact will help you judge how effective and appropriate your humor will be. If you have any doubts about how your humor will be received by your audience, don't use it.

Oral Presentation Rule #11: Do appropriate follow-up after the presentation.

After the presentation, promptly answer any questions that you were asked during the presentation. You can impress your customers by really meaning "I'll get

TABLE 11–2: Gestures to Avoid and Use

Don't	Do
• Keep your hands in your pockets • Rustle change and keys in your pockets (remove change and keys from your pockets before you deliver the presentation) • Tap a pencil, pen, or pointer • Frequently scratch • Slouch over a lectern • Shift from foot to foot • Pace and turn uncontrollably • Turn your back toward the audience • Fumble the handling of your presentation graphics, props, or equipment	• Stand straight, without constantly leaning on or gripping the lectern • Use your fingers and hands to emphasize major points • Use a pointer device to focus on key points on the screen • Use your face to accent emotion and feelings in the presentation • Occasionally step toward the listeners to decrease distance between you and them • Maintain frequent eye contact with the audience

back to you on that." Also, if allowed by customer contact rules during the selection process, on the day after your oral presentation, send the customer a letter to show your appreciation for the presentation opportunity and to summarize key presentation points.

Rules for Making and Using Oral Presentation Graphics

> A visual serves one main purpose: to help make a point. This concept often gets forgotten, and charts are tossed into the presentation because they're there. It's better to figure out the message and then determine the best way to show that. Many visuals have been wisely eliminated or extensively modified by the question, "What point is this visual intended to make?"
>
> (Leech, 1993, p. 139)

As previously noted, make graphics the backbone of your oral presentation. Because of the importance of graphics, this section provides more details about their development and use.

■ *Presentation Graphics Rule #1: Discover your listeners' preferences.*

Ask prospective listeners what graphics they prefer, applying the same principles of audience analysis that you used for the written proposal. When you ask, give your audience a choice of media, such as slides, overhead transparencies, projected computer graphics, and videotape.

■ *Presentation Graphics Rule #2: Prepare graphics when you prepare presentation content.*

Don't develop your presentation graphics as an afterthought. Instead, develop your graphic ideas as you prepare the content of your oral presentation. The storyboard approach used for the written proposal can be used to plan your graphics. Although the proposal can provide ideas for presentation graphics, a graphic that is effective in the proposal document may be too hard to read or understand as a visual displayed on a wall screen. Consider the use of graphic artists and speech coaches to help design the graphics. Give these professionals enough lead time to do their best work.

■ *Presentation Graphics Rule #3: Make graphics visible and understandable to your audience.*

The best graphics rely on visual images, not words. Keep them simple by avoiding the clutter of too much text. If you need to amplify a point, you can explain it orally during the presentation. When you need to put words in a visual—for example, by using a bulleted list of major points—pare them down to a minimum with single words or short phrases.

To work with the short-term memory of your audience, organize thoughts into groups of about four items. For example, if your written proposal mentions 10 main benefits of a new product, the oral presentation should distill this list to about four of the most important benefits. If you must mention all 10, you might cluster the points into three separate graphics.

The details in your graphics must be visible and understood by the audience in all parts of the presentation room. Learn about the room size and audience arrangement to ensure that the graphics details are appropriately sized. Also remember that the image size of a graphic projected on a wall or portable screen is affected by the distance between the projector and the screen. If possible, adjust the distance between the projector and screen for the best graphic display.

■ *Presentation Graphics Rule #4: Use colors carefully.*

Colors can add flair to visuals, but follow these simple guidelines to make colors work for you:

- Have a reason for using them, such as the need to highlight three different bars on a graph with three distinct colors.
- Use only dark, easily seen colors, and be sure that each color contrasts well with its background and with each other. For example, yellow text on white would be hard to read.
- Use no more than three or four colors in each graphic to avoid a confused effect.
- For variety, consider using white on a black or dark green background.

Presentation Graphics Rule #5: Leave graphics up long enough, but not too long.

How long is long enough? Oral Presentation Rule #2 recommends displaying graphics for about two minutes each. However, graphics should be shown only while you speak to the applicable point(s). For example, as you display a graph with an overhead transparency, you could say, "As you can see from the graph, the projected revenue increases until it reaches its maximum in two years." You then could pause, and leave the graph displayed a bit longer for the audience to absorb your point.

How long is too long? A graphic outlives its usefulness when it remains in sight after you have moved on to another topic. This error invites the audience to study the graphic and ignore what you're now saying. If you use a graphic once and plan to return to it, take it down after its first use and show it again later.

The effective display of graphics requires you to control the timing of your presentation. You can lose control if you give audience members copies of your graphics for their use during the presentation. The members can move through these handouts at their own pace, instead of following the pace you're trying to maintain. However, handouts are appropriate when

- Your audience requests them
- You think no other visuals will do
- You want the audience to take notes on the handouts during your presentation
- You distribute them as reference material after the oral presentation

Presentation Graphics Rule #6: While using graphics, maintain frequent eye contact with the audience.

Don't look at your graphics at the expense of frequent eye contact with your audience. Maintain a connection with your audience by looking back and forth from the visual to the audience faces. If you point to the displayed graphic, use the hand closest to the visual to keep the front of your body open to the audience. Using the opposite hand can cause you to cross your arm over your torso and to turn your neck and head away from the audience.

You may find it easier to look at your audience if you let someone else control the display transition of your graphics. By using a simple nod or verbal command, you can prompt your assistant to display the next visual. However, you might be more comfortable by controlling your graphics. If you do maintain control, avoid getting preoccupied with the operation of audiovisual equipment or electronic control devices.

Presentation Graphics Rule #7: Use graphics in your practice sessions.

Presentation practices should include the use of every planned graphic in its final form. Running through a dress rehearsal of your presentation without graphics would be much like a dress rehearsal for a play without costumes and props—you would be leaving out the parts that require the greatest degree of timing and orchestration. Practicing with graphics helps you improve transitions between graphics whether you'll be the only speaker or part of a team presentation.

■ *Presentation Graphics Rule #8: If possible, use your own equipment.*

Depending on someone else's audiovisual equipment comes with risk: new bulbs with a 100-hour life decide to blow and there are no extra bulbs, your laptop computer doesn't work with the LCD projector provided, the outlet near the projector doesn't work, the wall screen won't stay down, and the extension cords are defective, too short, or missing—the list could go on. Even if the equipment works, you may find that it operates differently from your equipment.

Avoid these problems by using your own equipment and testing it in the presentation area well before the presentation begins. If you don't own the required equipment, consider leasing it.

■ *Presentation Graphics Rule #9: When you don't use your own equipment, do backup planning.*

Be prepared for the worst when you have to rely on someone else's audiovisual equipment. Here are a few ways to ward off disaster:

- Before the presentation, contact the people who'll be responsible for providing the equipment. Ensure that they have equipment you need for the time you need it. Get familiar with the operation of equipment well before your presentation starts.
- Take easy-to-carry backup supplies with you, such as an extension cord, an overhead projector bulb, a flip chart, felt tip markers, or chalk.
- If you plan to use projected computer graphics, consider bringing an overhead projector and transparencies as a replacement if the computer or its projector fails.
- Bring printed copies of your graphics to hand out as a last resort.

Never be in position of having to apologize for equipment that's broken or missing, or that you don't know how to operate.

Tips for Overcoming Nervousness

The problem of nervousness deserves special mention, because it's so common among speakers. Virtually everyone who gives speeches feels some degree of nervousness before "the event." An instinctive "fight or flight" response can kick in for people who have an absolute dread of presentations. In fact, surveys have determined that many rate public speaking at the top of their list of fears, even above sickness and death! Given this common response, let's examine the problem and some suggestions for overcoming—or at least controlling—nervousness.

Most of us feel comfortable with informal conversations in which we can voice our views to friends and indulge in impromptu exchanges. We're used to this casual exchange of ideas. Formal presentations, however, put us into a more structured, more awkward, and thus more stressful environment. Despite the fact that we may know the audience members are friendly and interested in our success, the formal context triggers nervousness that's sometimes difficult to control.

The nervous response is normal and, to some degree, useful because it gets you "up" for the speech. Adrenaline pumping through your body can generate a degree of enthusiasm that propels the presentation forward and creates a lively performance. Just as veteran actors admit to some nervousness helping improve their performance, speakers can benefit from the same effect. The problem occurs when nervousness becomes so overwhelming that it affects the presentation quality.

As the cliché goes, don't try to eliminate "butterflies" before a presentation—just get them to fly in formation. Acknowledge that a certain degree of nervousness will always remain. The following are suggestions to ease your nerves.

■ *No Nerves Tip #1: Know your presentation content.*

The most obvious suggestion is also the most important one. If you prepare your presentation well, your command of the material will help reduce your anxiety—especially at the start of the presentation when nervousness is usually at its peak. Be so sure of the presentation content and graphic details that your listeners will overlook any initial discomfort you feel.

■ *No Nerves Tip #2: Prepare yourself physically.*

Your physical well-being before the speech can have a direct bearing on your anxiety. There's a connection between your mental and physical well-being, suggesting that you should take the following precautions before your presentation:

- Avoid caffeine or alcohol for at least several hours before you speak. You don't need the additional jitters brought on by caffeine or the false sense of ease brought on by alcohol.
- Eat a light, well-balanced meal within a few hours of speaking. However, don't overdo it—particularly if a meal comes right before your speech. If you're convinced that eating will increase your anxiety, wait to eat until after speaking.
- Practice deep-breathing exercises before you speak. Inhale and exhale slowly, making your body slow down to a pace you can control. If you can control your breathing, you can probably control nervousness.
- Exercise normally on the day of the presentation. A good walk can help invigorate you and reduce nervousness. However, don't wear yourself out by exercising more than you would normally.

■ *No Nerves Tip #3: Picture yourself giving a great presentation.*

Speakers can become nervous because their imaginations are working overtime. Instead of images of success, they bombard their psyches with images of failure.

To avoid this negative thinking, mentally see yourself successfully taking the following steps for the presentation:

1. Arriving at the room
2. Feeling comfortable at your chair
3. Getting encouraging looks from the audience
4. Giving an attention-getting introduction
5. Presenting your supporting points with clarity and smoothness
6. Ending with an effective wrap-up
7. Fielding questions with confidence

Called "imaging," this technique helps program success into your thinking and helps control negative feelings that can pass through the minds of even the best speakers.

■ *No Nerves Tip #4: Arrange the room as you want.*

To control your anxiety and to feel at ease, assert at least some control over the physical environment. Look for something that you can adjust to meet your preference. For example, have the positions of the audience chairs and the speaker lectern arranged to your satisfaction, and arrange for adequate lighting. It's a matter of your asserting control, so that your overall confidence is increased.

■ *No Nerves Tip #5: Have water nearby.*

Extreme thirst and dry throat are physical symptoms of nervousness that hurt the delivery quality of an oral presentation. To avoid these symptoms, have water available during your presentation. Prepare for this need ahead of time, so that you don't have to interrupt your presentation to pour the water.

■ *No Nerves Tip #6: Engage in casual banter before the speech.*

If you have the opportunity, chat with members of the audience before the speech. This ice-breaking technique can reduce your nervousness and help you establish rapport with the audience. It can also give you a "friendly" face to look at as you speak.

■ *No Nerves Tip #7: Remember that you're the expert.*

As a "psyching up" exercise before you speak, remind yourself that you'll be speaking on a topic about which you have useful knowledge. Expect your listeners to be eager about receiving this information. Tell yourself, "I'm the expert here!", and talk and act like one.

■ *No Nerves Tip #8: Don't admit nervousness to the audience.*

No matter how anxious you feel, never admit it to others. First, you shouldn't want listeners feeling sorry for you. Second, nervousness isn't always apparent to the audience. Your heart may be pounding, your knees shaking, and your throat dry, but the audience members may not note these symptoms. Why draw attention to the problem by admitting to it? Third, you can best defeat initial anxiety by simply pushing right on through it, concentrating on the task at hand.

■ *No Nerves Tip #9: Pace yourself.*

Feeling nervous can lead speakers to proceed too quickly through their presentations. This problem is less likely to occur if you prepare and practice well and critique the recordings of your practice. Video and audio recordings of your practice can help you detect if your pace is too quick (or too slow) and then lead you to adjust the pace as required. As you speak, constantly remind yourself to maintain an appropriate pace. If you've had a problem of speeding through your presentations, you might even write "Slow down!" in the margin of your notes.

■ *No Nerves Tip #10: Join a speaking organization.*

The previous guidelines will help you reduce your anxiety about a specific oral presentation. To help solve the problem over the long term, however, consider joining an organization, such as Toastmasters International, which promotes the speaking skills of its members and has chapters that meet at many companies and campuses. These meetings provide an excellent, supportive environment in which members can refine their speaking skills.

NEGOTIATING

> To become a successful negotiator you must discard a few human characteristics. Everyone, at one time or another, enjoys a position of power over another person and has the capability to impose desires, wishes, or wants upon that person. This imposition of power is usually displayed in the form of force, intimidation, or just plain old bluffing. This is not what successful negotiation is all about. Such actions may gain your immediate objective, but somewhere down the road the fiddler must be paid. . . . Successful negotiating is working together to achieve a common goal satisfactory to both parties.
>
> (Goodowens, 1989, p. 354)

As a member of a proposal team, you may not be involved in negotiating the final terms of a contract based on the proposal you wrote. Instead, contract negotiation may be a specialized task of contracts, legal, or sales organizations in your company. Regardless of your involvement, you should have a basic understanding of how effective negotiation can turn proposals into contracts.

We negotiate almost every day of our lives. By choice or chance, we find ourselves in give-and-take discussions to negotiate issues as diverse as major and minor purchases, relationships with spouses or friends, performance evaluations with bosses or employees, and proposal details with customers or in-house managers.

Rules for Negotiating

In the past, the negotiation process could be characterized by trickery, intimidation, and manipulation—resulting in "winners" and "losers" and lots of warlike imagery. Participants could be seen as battlefield adversaries, taking up extreme positions, defending and attacking each other's flanks, finally agreeing reluctantly to a middle ground, and then departing wounded, claiming victory but really uncertain of who had won the battle.

However, there is now less emphasis on the war-zone approach that demands an "I win, you lose" mentality. As a negotiator, you should enter the process searching for common ground for a very practical reason: long-term relationships are at stake. In later negotiations, you're much more likely to achieve success if the present negotiation helps both parties. This goal—"we both win"—can be achieved by following negotiation rules that encourage real communication rather than a pitched battle. The following rules will help you weave negotiation skills into your proposal writing, contract negotiation, and personal activities.

■ *Negotiation Rule #1: Think long term.*

Enter every negotiation with a long-term strategy in which your goal is to maintain and nurture the continuing relationship with the person on the other side. In business, your concern should be the repeat business that may accrue when customers believe a previous contract with you has treated them fairly. The value of repeat business far outweighs a few more dollars on one job. Thinking long term, then, is an attitude that you should project into your proposal and negotiation process. The remaining rules for negotiating offer specific techniques for reaching that long-term goal of building relationships.

■ *Negotiation Rule #2: Prepare a negotiation strategy.*

If you're to be involved in contract negotiation with the customer, prepare by fully knowing the proposal content and having a range of terms, conditions, and prices you can offer the customer. Document and review your negotiation strategy before the negotiation begins. Knowledge of your proposal will help you discuss negotiation points that are involved with work scope or cost details. If it's a complex proposal, you might need the assistance of technical or management specialists during negotiation. With a range of acceptable terms, conditions, and prices, you enter the negotiation ready to give your customer choices. The availability of customer choices—or options—is covered in the next rule.

■ *Negotiation Rule #3: Explore many options.*

The negotiating process often begins with the consideration of only two outcomes—that is, the specific objectives of the two negotiating parties. You can escape this trap by exploring many options in the early stages, which then can lead to reaching creative solutions that satisfy both sides. The point is that examining many possibilities keeps both of you talking and getting to know each other's broad concerns, rather than solidifying your separate positions.

You can begin the process of offering options in your proposal. For example, your proposal might offer several design features of your product, ranked by a cost-benefits analysis you've performed. During contract negotiation, the customer can then choose the preferred approach.

■ *Negotiation Rule #4: Find the shared interests.*

If you succeed in keeping the options open during a negotiation, you can begin to discover points on which you agree. Draw attention to these points of agreement rather than the points of conflict. Finding shared interests helps establish a friendship, which in turn makes both sides more willing to reach consensus.

■ *Negotiation Rule #5: Listen carefully.*

Despite multiple options and shared interests, negotiations usually return to basic differences. An effective technique at this point is to focus on the rationale behind your counterpart's views, not the views themselves. It can help the negotiation—and the support of your own view—to ask questions and then listen carefully to the answers coming from the other side.

Asking probing questions about your customer's rationale benefits both you and the entire negotiation because you

- Give your customers the opportunity to explain their views, thus breaking out of the attack-counterattack cycle
- Discover what motivates the customer, making it more likely that you'll find an appropriate response and reach consensus
- Expose weak logic and unsupported demands of your customer
- Move closer to objective standards on which to base negotiations

From your persistent questioning, careful listening, and occasional responses, information may emerge that otherwise would have remained buried.

■ *Negotiation Rule #6: Be patient.*

Hard-sell negotiations can push participants to quick decisions, often to the regret of at least one of the parties. The better approach is to slow down the negotiation process to prevent decisions from being based on the emotionalism of the moment. Don't let your emotions overtake objectivity in any negotiation. Good negotiated settlements should stand the test of time. When one party feels pressured, mistakes can be made. Well-thought-out decisions are more likely to produce better long-term relationships, which should be a major goal of your negotiations.

■ *Negotiation Rule #7: Look back.*

Some may believe that, once an agreement is negotiated, you shouldn't look back to second-guess yourself, because it'll only make you less satisfied with something that can't be changed. This attitude mistakenly assumes that negotiations are spontaneous phenomena that can't be analyzed. If you've conducted your negotiations methodically, you'll have much to gain from postmortems—particularly if you document your negotiation strategy, process, and results.

Take notes during negotiations, and document the process and results in a journal or report. Use this documentation to develop lessons learned to prepare for future negotiations, particularly with the same customer. In the journal or report, answer the following type of questions:

- What options were explored before a decision was made?
- What shared interests were discovered?
- Did you emphasize these shared interests?
- What questions did you ask?
- How did you show that you were listening to responses?

Analyze every negotiation to discover what went right and wrong during the proceedings. As with other communication skills, such as writing and speaking, the ability to negotiate should improve with use.

Strategies for a Best and Final Offer

Before customers decide to negotiate a contract based on competing proposals, they may start a form of price negotiation by asking for a best and final offer (BAFO) bid from the competing bidders. This step allows you to bid one last price before the award decision is made. The normal response to a BAFO request is to lower the price.

Although the BAFO request occurs after proposal submittal, begin to develop a BAFO strategy before the proposal is ever delivered. Before proposal submittal, be alert for any customer indications that a BAFO request will be made. However, customers may be careful to avoid any hint of a forthcoming BAFO, because they don't want bidders to offer an artificially high first price with the plan of lowering it later. Use the questions in Table 11–3 to plan your BAFO strategy.

TABLE 11–3: Questions for Planning a BAFO Strategy

Topic	Question
BAFO decision	Will you submit a final price with the original proposal, whether or not the customer asks for a BAFO?
Price decision	Will you submit a new price if the customer does ask for a BAFO?
Competitor assessment	How do you think your competition will respond to a BAFO request, and how will this affect your BAFO?
Competitive range	Is your current price within the competitive range that you think the customer will accept, and will you need a BAFO to stay in the competitive range?
BAFO justification	How will you justify a BAFO that lowers the bid price? For example, will you lower your profit margin, change a schedule for better labor efficiency, use a vendor for its lower-priced subcontract labor or materials, lower labor estimates based on new insight into customer needs, or offer to absorb costs for research that might otherwise be charged to the customer?
BAFO credibility	Will your new price seem reasonable and credible and not indicate that your first price was noncompetitive or bloated?
Strategy compatibility	Is your BAFO strategy compatible with your strategy planning for the original proposal? If not, are you convinced that the BAFO strategy is better?
Other factors	How will your BAFO strategy be affected by RFP amendments and question-and-answer and oral presentation sessions with the customer?
Long-term impact	How will your BAFO affect your future chances of getting more work with the same customer?
Proposal impact	Will the content of your proposal have to be revised to reflect labor or material cost changes that were made to reach the BAFO bid? How and when will the customer ask that these proposal changes be made?

POST-AWARD RULES

After the award decision is made, there are still some important steps to take in the proposal process. These activities are summarized in Figure 11–2. Note that most of these steps are performed whether your proposal is a winner or a loser. Use the following rules to perform important post-award actions.

■ *Post-award Rule #1: Ask the customer for a proposal debrief.*

> The debrief is perhaps the most useful meeting you will ever have with a client, at least with a client whose business you have just lost. A debrief is a meeting, common in most industries, that is given to you as a matter of courtesy to give you some insight into why you were unsuccessful in winning that client's business. It is an opportunity to learn the *real* reasons why you lost the business.
>
> (McCann, 1995, p. 253)

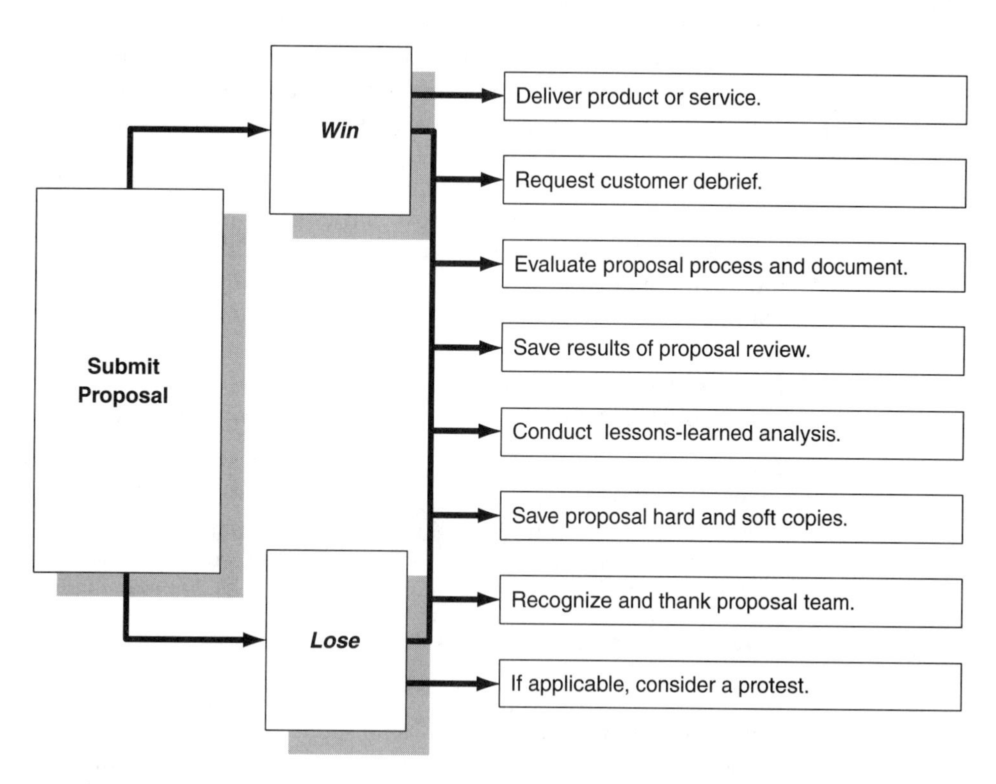

FIGURE 11–2
Post-award activities

Win or lose with your commercial or government proposals, ask the customer for an oral or written debrief about your proposal. Regardless of the debrief method, ground rules, or scope, your focus should be (1) to learn what you did well in the proposal so you can do it again in other proposals, and (2) to learn what you didn't do well, so you don't repeat it. The debrief isn't for debating or arguing about how your proposal was evaluated. Come to the debrief prepared to ask questions and take notes that will lead to feedback for the process described in Post-award Rule #4.

In the commercial world, the customer decides if a debrief is to be held and what it will cover. In contrast, if you had a losing proposal for a U.S. government contract that was not awarded based on the lowest price, you are entitled to a debrief if you ask for it via a written request to the contracting officer. Although the depth of U.S. government debrief can vary by agency, expect the debrief to provide feedback at least on the strengths and weaknesses of your proposal. Don't expect it to provide specific grades and details about your competition's proposals. See the Federal Acquisition Regulation (FAR) for the latest debrief rules.

Win or lose based on the award decision, also consider sending a questionnaire survey to your commercial customer. In this survey, ask about the effectiveness of specific proposal features. This survey could lead to or support a face-to-face debrief between you and that customer.

You can also gain valuable post-award information by asking for local, state, or federal government customer records available for public disclosure. For example, access to federal government records is authorized by the Freedom of Information Act (FOIA). Among the many possibilities, federal information can include contractual terms and proposal records about a contract for which your proposal was unsuccessful. The FOIA doesn't allow you to obtain proprietary information (competition sensitive) about your winning contractor competition. However, you may find information that can help you in the future to propose more effectively to the same or another government customer or to compete with the successful contractor. (If you send proposals for business with local and state governments, learn about your legal rights to obtain public information from these government customers.)

■ *Post-award Rule #2: Conduct an internal evaluation of your proposal process and document.*

> A post-mortem on the proposal after the customer's decision will always yield useful information. This is true whether the proposal won or lost. You can seek feedback from your customer, from your resource people, and from your own analysis. . . . There's no point in making a perfectly good mistake and then not learning from it.
>
> (Svoboda and Godfrey,1989, p. 131)

Win or lose, have your proposal staff identify the strengths and weaknesses of its proposal processes and document. Also include the proposal staff of your subcontractors or team companies in this evaluation. There are many points you could examine. For example, were your various proposal management plans effective?

If they weren't, why not? What obstacles did the proposal team face and how well did it overcome them? What were the most and least effective parts of the proposal and why?

■ *Post-award Rule #3: Save review comments and recommendations.*

Proposal reviews not only help you improve the proposal being reviewed but they also can help the quality of future proposals. Therefore, save the comments and recommendations generated by proposal reviews, including copies of the debrief materials and comment forms. It also would be helpful to annotate the review recommendations with the proposal team's response to these recommendations. The review information can help you identify (1) proposal mistakes you don't want to repeat and (2) effective proposal approaches that you do.

■ *Post-award Rule #4: Recommend improvements based on lessons-learned analysis.*

Perform lessons-learned analysis on the information described in the preceding three rules. Translate these lessons learned into recommendations for improving your proposal process, which includes the preparation and use of the 10 proposal management plans described in Chapter 2. Your ultimate goal should be to improve the effectiveness of your future proposals.

Analyze information about your proposal performance and costs by using statistical or numerical measurement of this analysis—also called metrics. The following are just a few examples of metrics that could be useful in lessons-learned development:

- Proposal cost and process metrics—bid and proposal (B&P) cost per proposal page (total labor, material, and travel costs to produce the proposal divided by proposal pages), breakout of actual proposal team work hours by specific proposal tasks, and B&P costs as a percentage of the proposed contract value for proposals that win and lose
- Discrepancy trend metrics—categorical listings and frequency of specific proposal discrepancies noted in internal reviews and customer questions and debriefs
- Performance metrics—win percentage (number of proposals won divided by number of proposals submitted) when your firm is a prime contractor, subcontractor, or an incumbent contractor (performing on a current contract up for rebid); or the win percentages according to the target customer, product offered, or geographical location of the target customer

Establish a quality control system that reviews, approves, and implements the recommended improvements. Also, ensure that the lessons-learned analysis, metrics information, and resulting recommendations are available for use by your future proposal teams.

Post-award Rule #5: Save hard and soft copies of the proposal document.

Save printed copies of the proposal as an information source for other proposals. If the proposal was a winner, copies of it should be distributed to those assigned to work the contract. In addition, save software files of the proposal text and graphics; you may be able to use their content with little or no revision in other proposals and, as a result, reduce the time needed to develop those proposals.

Post-award Rule #6: Protect your rights with the government customers.

> . . . protest procedures are very legalistic and can result in expenditure of great sums of money if attorneys are involved. First try to convince the contracting officer to change his way of conducting the solicitation. If you can't do that, decide whether the gains to be made if you win the protest outweigh the costs of pursuing the protest. In other words, make a business decision: "even if I win the protest, am I going to lose my shirt?" If you decide it is worthwhile to pursue the protest, do so. The right to protest is a powerful one, and contracting officers do not take it lightly.
>
> (McVay, 1987, pp. 157–158)

Know your rights when you're bidding for a government contract. Learn what recourse you have if you think the bid process wasn't properly executed by a prospective local, state, or federal customer. If you lose a proposal competition and think you weren't treated unfairly, you may have grounds for a formal protest. However, protests should be filed only after careful thought.

As an example, let's consider a protest to the federal government over government work solicited through an RFP (which by definition leads to a negotiated contract). A contractor may file this kind of protest (1) when it believes its ability to compete for the contract was hindered by a contracting officer who didn't follow all the proper procedures or (2) when it believes that the winning contractor didn't meet the requirements of the RFP. The protest is filed to the General Accounting Office (GAO), not the respective contracting officer. If a protest is upheld by the GAO, the contract awardee may be changed or the proposal evaluation process repeated; in either case, you could gain the contract. If it's upheld, you may also be reimbursed by the government for your costs of processing the protest and developing the proposal.

However, most protests regarding negotiated federal contracts aren't successful. Whether the protest succeeds or fails, your protest will require administrative attention by you and the government. This takes time and personnel resources. Just the filing of a protest can delay contract award, suspend an awarded contract, or stop an awarded contract—events that probably won't endear you to the contracting officer and affected agency.

The following are some key questions you should answer before proceeding with a protest:

- How confident are you that the protest will be upheld? If you don't strongly believe you have a winnable protest, it's probably best not to file.
- Does the prospect of winning the contract justify your cost of processing the protest?
- Will the protest—whether it's upheld or denied—jeopardize your chances of gaining other work with the same government customer in the future?
- If you win the contract after having your protest upheld, can you expect to have a cooperative relationship with the government customer? If the answer is no, will your ability to meet the contractual requirements be adversely affected?

A protest is a serious step, but one that can help protect your rights. For the most current guidelines about submitting and processing a protest to the federal government, refer to the latest FAR.

Post-award Rule #7: Thank the proposal team.

Proposal work is physically and mentally demanding. It often requires long hours, seven-day workweeks, and a work ethic that forces people to neglect themselves and their family and friends, at least temporarily. Win or lose, a company should thank the proposal team for its effort.

The thanks should be sincere, meaningful, and timely to the recipients. It can take the form of proposal team T-shirts or coffee mugs, framed copies of the proposal cover with a personal thank-you signed by the company president, or personal letters of thanks signed by higher level proposal or company management with copies to a person's employee file. Or appreciation can be shown with compensated time off, a victory celebration party, or even a party to lift spirits after a defeat. Avoid gestures such as sending superficial form letters of thanks to all participants on a distribution list or limiting the thanks to a regular staff meeting in which the gathered masses are perfunctorily thanked for their hard work.

Done properly, a thank-you and related recognition can boost morale and make proposal work more attractive to the people who'll be assigned to future proposal teams. It should make the proposal staff feel as special and important as they really are.

Post-award Rule #8: As a winner, deliver the product or service as required.

It's important to perform post-award analysis to improve the competitiveness of your proposals. But one of the best ways to gain new and repeat business is to deliver what was agreed to in the contract based on your proposal. Your past performance or related experience, often required content in a proposal, can be a key factor in the evaluation of your proposal by customers—especially U.S. government customers. If you have a proven record for delivering what you promise, it can be a strong indication that you can repeat this performance in the future.

SUMMARY

The proposal process continues even after the proposal is delivered.

You may need to respond to questions from the customer about your proposal. Use the following rules to respond to these questions:

1. Have an established contact point for the customer.
2. Assign responsibility and a deadline for the response.
3. Answer completely and tactfully.
4. Establish a review and approval process for all answers.
5. Provide answers to meet customer submittal instructions.

Customers or in-house managers may want you to make an oral presentation after the proposal is submitted. Use the following rules to prepare and deliver your oral presentations:

1. Use the Preacher's Maxim.
2. Plan your presentation to stay within the allotted time.
3. Use notes.
4. Practice, practice, practice.
5. Speak vigorously and deliberately.
6. Avoid those awful fillers.
7. Use rhetorical questions.
8. Maintain frequent eye contact.
9. Use appropriate gestures and posture.
10. Use humor carefully.
11. Do appropriate follow-up after the presentation.

Graphics are crucial to the success of an oral presentation. Use the following rules to develop presentation graphics:

1. Discover your listeners' preferences.
2. Prepare graphics when you prepare presentation content.
3. Make graphics visible and understandable to your audience.
4. Use colors carefully.
5. Leave graphics up long enough, but not too long.
6. While using graphics, maintain frequent eye contact with the audience.
7. Use graphics in your practice sessions.
8. If possible, use your own equipment.
9. When you don't use your own equipment, do backup planning.

Nervousness can be your biggest challenge during an oral presentation. Use the following tips to minimize your nervousness:

1. Know your presentation content.
2. Prepare yourself physically.
3. Picture yourself giving a great presentation.
4. Arrange the room as you want.

5. Have water nearby.
6. Engage in casual banter before the speech.
7. Remember that you're the expert.
8. Don't admit nervousness to the audience.
9. Pace yourself.
10. Join a speaking organization.

Negotiate to reach an agreement that all parties can accept. Use the following rules for effective negotiation:

1. Think long term.
2. Prepare a negotiation strategy.
3. Explore many options.
4. Find the shared interests.
5. Listen carefully.
6. Be patient.
7. Look back.

A best and final offer (BAFO) can be considered a form of price negotiation. Although a BAFO is given to the customer after proposal submittal, begin to develop your BAFO strategy even before you deliver the proposal.

Work remains to be done even with the announcement of the proposal winner. Use the following post-award rules to continue the proposal process:

1. Ask the customer for a proposal debrief.
2. Conduct an internal evaluation of your proposal process and document.
3. Save review comments and recommendations.
4. Recommend improvements based on lessons-learned analysis.
5. Save hard and soft copies of the proposal document.
6. Protect your rights with the government customers.
7. Thank the proposal team.
8. As a winner, deliver the product or service as required.

EXERCISES

1. Oral Presentation: In-House Proposal

Select a nonproprietary in-house proposal you have written during your work experience or one you wrote in response to a Chapter 8 exercise. Prepare an oral presentation (5 to 10 minutes long) that could serve as an overview of the chosen proposal. Include two to five graphics. Assume that this in-house "customer" has expressed initial interest in your proposal and now wants a brief oral overview.

2. Oral Presentation: Sales Proposal

Select a nonproprietary sales proposal you have written during your own work experience or one you wrote in response to a Chapter 7 or Chapter 8 exercise. Prepare an oral presentation (5 to 10 minutes long) that could serve as an overview of the chosen proposal. Include two to five graphics. Assume that this customer has expressed initial interest in your proposal and now wants a brief oral overview.

3. *Oral Presentation: Professional Conference*

Select a product or service with which you are, or can become, familiar—either through experience, academic course work, or research. Assume that you represent a company that sells this product or service. Also assume, also, that you've been asked to deliver a presentation (10 to 15 minutes long) on the product or service, with five to seven graphics. Your presentation is one of many to be given at an annual conference sponsored by a professional organization in your field. The speech should be largely informative, as opposed to a sales pitch; however, keep in mind that some members of the audience may be prospective customers. In the oral presentation, don't include proprietary information of any organization.

4. *Negotiation Analysis: Actual Case*

Evaluate a negotiation process that you've experienced personally. Write a report or deliver an oral presentation describing your level of success and the way you followed this chapter's negotiation rules. In the report, don't include proprietary information of any organization.

5. *Negotiation: Videotaped Role-Playing*

For this exercise, perform a simulated negotiation and then evaluate the exercise by viewing a videotape of it. Follow these guidelines:

- Choose a partner and assume that the two of you are competing against each other in a negotiation.
- Select a nonproprietary subject on which the two of you have gained information, either from research or work experience.
- Select an evaluator who'll observe the negotiation and keep notes on the process.
- Submit your written objective(s) for the negotiation to the evaluator before the session and have your counterpart do the same.
- Arrange to have the session videotaped.
- Perform a role-playing exercise for 15 to 30 minutes.
- Have the evaluator analyze the manner in which you both followed the chapter's negotiation rules, in light of the written objectives you submitted beforehand.
- Review the videotape with your counterpart and the evaluator.

6. *Lessons-Learned Analysis: Written Class Projects*

What mistakes have you made during the planning, managing, writing, editing, or production of the proposal documents you have produced for this class? Translate these lessons learned to recommendations you might implement to write a formal sales proposal in response to an RFP solicitation. Present your analysis and recommendations in a group discussion, an individual oral presentation, or an individual written report.

7. Lessons-Learned Analysis: Organizational Improvement Process

Based on your personal experience in any organization, describe what nonproprietary mechanisms and procedures that organization used to monitor its performance and to implement changes for improving its operation. Explain how well these approaches worked and how they could be applied by a company's proposal organization. Present your analysis and observations in a group discussion, an individual oral presentation, or an individual written report.

8. Response to Customer Questions

Discuss how you might "read between the lines" in questions received from the customer about your proposal. Also discuss how this reading of the questions could affect your response to the customer. Conduct this exercise as a class discussion.

9. Question Development: Sales Proposal

Divide the class into teams. After reviewing the formal sales proposal model in Chapter 7, have each team develop a list of 10 questions to clarify the proposal or to ask for useful information that wasn't in the proposal. Have each team identify the questions. Compare the variety of questions by discussing the content and tone of the questions developed by each team.

Enclosure (1) Bibliography

Bowman, Joel P., and Bernadine P. Branchaw. 1992. *How to write proposals that produce.* Phoenix, AZ: The Oryx Press.

Burke, Clifford. 1990. *Type from the desktop: Designing with type and your computer.* Chapel Hill, NC: Ventana Press.

Clauser, Jerome K. 1993. "The importance of proposal illustrations." In *How to create and present successful government proposals: Techniques for today's tough economy,* ed. James W. Hill and Timothy Whalen, pp. 226–235. Piscataway, NJ: IEEE Press.

Edmunds, Don L. 1993. *Source selection: A seller's perspective, how the federal government selects contractors & products.* 2nd ed. Vienna, VA: Holbrook & Kellogg.

Franklin Quest Consulting Group (formerly Shipley Associates). 1992–1996. *Writing winning proposals: Capturing commercial business workshop manual.* Salt Lake City, UT: Franklin Quest Co.

Freeman, Lawrence H., and Terry R. Bacon. 1990, reprinted 1995. *Style guide: Revised edition.* Bountiful, UT: Shipley Associates.

Geever, Jane C., and Patricia McNeill. 1993. *The Foundation Center's guide to proposal writing.* New York: The Foundation Center.

Gilpatrick, Eleanor. 1989. *Grants for nonprofit organizations: A guide to funding and grant writing.* New York: Praeger Publishers.

Goodowens, James B. 1989. *A user's guide to federal architect-engineer contracts.* New York: The American Society of Civil Engineers.

Hansen, Robert M. 1992. *Winning strategies for capturing defense contracts.* Arlington, VA: Gloria Magnus Publishing.

Helgeson, Donald V. 1994. *Engineer's and manager's guide to winning proposals.* Norwood, MA: Artech House.

Hill, James W. 1987. "How to write a good executive summary for a proposal." In *How to create and present successful government proposals: Techniques for today's tough economy,* ed. James W. Hill and Timothy Whalen, pp. 166–173. Piscataway, NJ: IEEE Press.

Hill, James W. 1993. "A potpourri of proposal management considerations." In *How to create and present successful government proposals: Techniques for today's tough economy,* ed. James W. Hill and Timothy Whalen, pp. 115–124. Piscataway, NJ: IEEE Press.

Holtz, Herman. 1986. *The consultant's guide to proposal writing: How to satisfy your clients and double your income.* New York: John Wiley & Sons.

Jacobs, Daniel M., Janice M. Menker, and Chester P. Shinaman. 1990. *Building a contract: Solicitations/bids and proposals, a team effort?* Vienna, VA: National Contract Management Association.

Leech, Thomas. 1993. *How to prepare, stage, & deliver winning presentations.* 2nd ed. New York: AMACOM, a division of American Management Association.

Luther, William M. 1992. *The marketing plan: How to prepare and implement it.* New York: AMACOM, a division of American Management Association.

McCann, Deiric. 1995. *Winning business proposals.* Dublin, Ireland: Oak Tree Press.

McVay, Barry L. 1987. *Getting started in federal contracting: A guide through the federal procurement maze.* 2nd ed. Burke, VA: Panoptic Enterprises.

Meador, Roy. 1991. *Guidelines for preparing proposals.* 2nd ed. Chelsea, MI: Lewis Publishers.

Miner, Lynn E., and Jerry Griffith. 1993. *Proposal planning and writing.* Phoenix, AZ: The Oryx Press.

Newman, Larry. 1994. *Developing capture plans workshop manual.* Bountiful, UT: Shipley Associates.

Nocerino, Joseph T. 1993. *How to prepare winning competitive proposals: A step-by-step guide for mastery and success.* 5th ed. Vienna, VA: Century Planning Associates.

O'Connor, John P. 1993. "The writer's responsibility for proposal graphics." In *How to create and present successful government proposals: Techniques for today's tough economy,* ed. James W. Hill and Timothy Whalen, pp. 236–254. Piscataway, NJ: IEEE Press.

Pfeiffer, William S. 1989. *Proposal writing: The art of friendly persuasion.* Columbus, OH: Merrill Publishing Co.

Pfeiffer, William S. 1997. *Technical writing: A practical approach.* 3rd ed. Upper Saddle River, NJ: Prentice-Hall.

Pfeiffer, William S. 2000. *Technical writing: A practical approach.* 4th ed. Upper Saddle River, NJ: Prentice-Hall.

Pugh, David G. 1993. "Managing the proposal effort." In *How to create and present successful government proposals: Techniques for today's tough economy,* ed. James W. Hill and Timothy Whalen, pp. 82–89. Piscataway, NJ: IEEE Press.

Sant, Tom. 1992. *Persuasive business proposals: Writing to win customers, clients, and contracts.* New York: AMACOM, a division of American Management Association.

Shipley Associates. 1995. *Writing winning proposals: Capturing federal business workshop manual.* Bountiful, UT: Franklin Quest Co.

Svoboda, Krasna, and Richard L. Godfrey. 1989. *The perfect proposal: A vendor's guide to award-winning telecommunications & data processing proposals.* New York: Telecom Library.

Tracey, James R. 1993. "STOP, GO, and the state of the art in proposal writing." In *How to create and present successful government proposals: Techniques for today's tough economy,* ed. James W. Hill and Timothy Whalen, pp. 51–81. Piscataway, NJ: IEEE Press.

Usrey, Nancy J. 1996. *Insider's guide to SF254/255 preparation.* 2nd ed. Natick, MA: Zweig White & Associates.

Weiss, Edmond H. 1990. *100 writing remedies: Practical exercises for technical writing.* Phoenix, AZ: The Oryx Press.

Whalen, Tim. 1996. *Writing and managing winning technical proposals.* 3rd ed. Vienna, VA: Holbrook & Kellogg.

Woolston, Donald C., Patricia A. Robinson, and Gisela Kutzbach. 1988, reprinted 1990. *Effective writing strategies for engineers and scientists.* Chelsea, MI: Lewis Publishers.

Enclosure (2) Exercise Solicitation and Evaluation Criteria

EXERCISE NOTES TO THE INSTRUCTOR

Have your students use the following request for proposal (RFP) to develop a team-written formal proposal. Fill in the RFP blanks to tailor it for your school and the scheduling of your class. The RFP requires the student teams to prepare for and attend a bidders' conference; develop and submit a formal sales proposal to meet a deadline; and, after proposal delivery, make an oral presentation to support the proposal. (Although the development of a cost proposal is noted in the RFP, students are not to submit a cost proposal for this exercise. Consider this RFP to be the first volume of a two-part RFP, the second volume being the future request for a cost proposal.)

The following are exercise recommendations for the instructor, who is to serve as the contract administrator identified in the RFP. (However, instructors are encouraged to change the content and use of the RFP and the grading criteria to best meet their training needs.)

- Assign five to seven students per team, with care to balance the skills among the teams. (It's assumed that you'll have at least two student teams—although the exercise can be tailored for one.) Have the teams organize themselves and choose their own proposal manager. Consider giving team proposal managers extra credit for assuming this responsibility.
- Require each team to develop and submit a proposal preparation schedule, strategy plan, requirements matrix, and proposal storyboards for your information and, if desired, for your review and comment.
- Limit the amount of money a team can spend on the production of the proposal, including the cost of binders, covers, tabs, graphics, and proposal reproduction. Focus this exercise on the content of the proposal, not on its production and packaging.

- Have each team provide you and any other team(s) with a copy of its proposal. Seeing how others responded to the same RFP can be a good learning experience for the students.
- As desired, amplify or revise the RFP instructions for the team oral presentations.
- Provide separate team grades for the proposal and the oral presentation, giving more weight to the proposal document grade. Make the team grade equal to the grade given to each student on that team (with the proposal manager receiving extra credit, if desired). It's recommended that you use the evaluation criteria in the RFP to develop the team proposal grade.
- During the oral presentations, have another instructor serve as the Professor's Academic Board (PAB) chairperson and ask questions about the team proposals and oral presentations. This instructor should review the RFP and all team proposals before attending the presentations. (You, as the class instructor, are to attend the oral presentations in the role of the contractor administrator.)
- Have each team evaluate the "competitor" proposal(s) based on the evaluation criteria and then give an oral presentation about its evaluation. This is a good way for students to practice their review and presentation skills and to learn from each other. In particular, the oral presentation will give the students practice in critiquing their peers—a task often required of proposal reviewers in the business world. It's recommended that you (1) grade the presentations, but not let any comments that a team may make influence the grade you give for a "competitor" proposal, and (2) stress to the students that their comments about a "competitor" proposal will in no way affect how you grade that proposal.

TO: Prospective Offeror (Bidder Team)

FROM: Contract Administrator

SUBJ: (School's name ______________________________) Request for Proposal (RFP) # ____ , Volume 1 of 2, Problem Analysis Study (PAS) for the (a selected academic degree program: ____).

DATE: ___________

1.0 STATEMENT OF WORK

(School name and location __________) will award a contract to perform a Problem Analysis Study (PAS) to identify improvements for the subject academic program (referred to as the "program" in this RFP).

1.1 PAS Evaluation Topics

The PAS, to be planned and performed by a Contractor Study Team (CST), competitively selected based on this RFP, will address the following issues/problems:

1.1.1 Course selection, content, and scheduling
1.1.2 Competitive recruitment of qualified students for the program
1.1.3 Relationship between the program and local businesses and industry
1.1.4 Other issues that are important to the future effectiveness and growth of the program

1.2 CST and (School) Coordination

The CST will coordinate its work with a yet-to-be-selected team of professors, known as the Professor's Academic Board (PAB). The offeror will recommend how the CST will integrate its activities and work with the PAB during the performance of the PAS contract, and will recommend how the PAS could be supported by school student organizations. The selected CST will be staffed only by the individuals who contributed to their corresponding proposal and are identified in that proposal as the proposed CST members.

1.3 PAS Task Description

The PAS project to be proposed will be divided into four tasks:

1.3.1 Task 1—Identify and research evaluation topics listed in SOW Section 1.1.
1.3.2 Task 2—Develop and submit a formal report that lists and ranks the importance of target problems. This report will be submitted to the PAB, which will review the report and select problems for further action.
1.3.3 Task 3—Develop potential solutions for the problems selected by the PAB during Task 2.
1.3.4 Task 4—Develop a final formal report that describes approaches for resolving the problems targeted by PAB selection. These recommendations will address specific approaches and resources needed to resolve the problems, and include such topics as the need for coordinated student and faculty activities and the estimated costs of implementing the recommendations.

1.4 PAS Schedule

The resulting contract, to be awarded on (date _______), will be funded for six months. In the proposal, the offeror will provide a schedule identifying key milestones within each of the four PAS task periods. The duration of each task period is up to the offeror; however, all four tasks must be completed within six months after contract award.

1.5 Reports (Deliverables), Meetings, and Reviews

The CST team will provide the PAB and the contracting administrator with monthly written status reports during the performance of all four PAS tasks. In addition, the Task 2 and 4 formal reports will be submitted to all PAB members. All status and formal reports will have a content and format approach proposed by the CST in the proposal. The approach will be approved by the contracting administrator during contract negotiation. In its proposal, the offeror will describe the proposed content of these reports, recommend meetings and reviews to be scheduled during all four PAS task periods, and identify facility and equipment requirements needed to support these meetings and reviews.

2.0 PROPOSAL INSTRUCTIONS

The offeror will develop and submit one volume: Technical & Management Proposal. Based on the contractor administrator evaluation of this proposal, a qualified offeror will be asked to submit a cost proposal. At that time, Volume 2 of this RFP will provide instructions for the cost proposal.

(_____ copies) of the Technical & Management Proposal will be submitted in individual three-ring binders to the contract administrator, at (location __________), no later than (time and date _______). The copies will include the original proposal, indicated as such and signed by all members of the proposal team, and _____ copies of the original. The proposal cover will at a minimum identify the offeror's name, the RFP number, and the date of the proposal. All proposal copies will be delivered securely wrapped in a box, which will have no markings on the outside except for the offeror's name.

No pricing for any offered product or service will be included in the Technical & Management Proposal. However, the proposal can address the cost-effectiveness of the proposed study approach.

A bidders' conference for all prospective offerors will be directed by the contract administrator beginning at (time, date, and location ______________________). At the conference in the presence of all attendees, the contract administrator will orally answer written questions submitted by all prospective offerors. These written questions accompanied by a transmittal letter will be submitted to the contract administrator at the bidders' conference. At the conference, the contract administrator reserves the right to answer only the written questions, choosing not to answer oral questions by those in attendance.

On (date _____) after proposal submittal, each offeror will provide the contract administrator and, if available, the PAB chairperson with an oral presentation summarizing its Technical & Management Proposal. The presentations will be held at (location ______________________).

In the main presentation, each offeror will present a summary of each of the four major sections of its written proposal: (1) Task Approach, (2) Problem Assessment, (3) Study Management, and (4) Personnel Resumes. In addition, a list of features and benefits for the offeror's approach will be provided as part of the presentation introduction. Each offeror will decide which member(s) of the proposed CST will participate in the oral presentation; however, all proposed offeror CST members will be required to attend the presentation and be available to answer questions. The offeror's main presentation will not exceed 15 minutes. Led by the contract administrator, a question-and-answer session about the offeror's proposal will follow the main presentation. If in attendance, the PAB chairperson will also participate in this question-and-answer activity.

2.1 Volume I Instructions

In the Technical & Management Proposal, the offeror will provide the following information in the specified four proposal sections:

Section 1. Executive Summary—Provide an overview of the proposed task and management approaches and the qualifications of the proposed CST members. (This section will not exceed four pages.)
Section 2. Task Approach—Describe the proposed approach to performing each of the four major PAS tasks. Include a functional flow diagram showing (1) the subtasks proposed to complete each of the four major PAS tasks, (2) the relationship of these subtasks within each major PAS task, and (3) the overall relationship of the four major PAS tasks.
Section 3. Problem Assessment—Provide a preliminary assessment of potential PAS topics, describing problems that could be addressed by the PAS and possible solutions for these problems. Include examples of potential problems and their solutions for all four categories of PAS evaluation topics identified in SOW Section 1.1 (subsections 1.1.1–1.1.4).
Section 4. Study Management—Describe the proposed CST organization and how it will manage the study. At a minimum, include a description of the

- Internal coordination of the CST and its interface and coordination with members of the PAB and student organization(s)
- CST management control of the PAS schedule and budget
- Development, submittal, review, and content of status and formal reports
- Project scheduling for all four PAS major tasks, including a schedule that clearly shows the activities and milestones within each major task, the expected beginning and ending dates of each major task, and the milestones for all required and planned management/deliverable activities, such as meetings, reviews, and the submittal of status reports and formal reports

Section 5. Personnel Résumés—Provide a résumé for each member of the proposed CST, describing the member's experience, skills, and qualifications for PAS participation. CST members will be required to be familiar with the program; have strong research, analytical, and problem-solving skills; have strong writing and oral presentation skills; and have the proven ability to work in a team environment. The offeror will describe the PAS role and responsibility of its proposed CST members.

Preceding Section 1 in the proposal will be a title page, table of contents, figure list, and compliance matrix (cross-referencing where RFP requirements are addressed in the proposal).

Volume I (Sections 1–5) will not exceed 40 pages (8 1/2″ × 11″ page size). Graphic foldouts (11″ × 17″ page size) will count as two pages. The proposal will be written with an 11 pt. serif font, single-spacing between lines in paragraphs and double spacing between paragraphs. The 8 1/2″ × 11″ pages will have 1″ side margins and a 1/2″ top and bottom margins. All pages, including foldouts, will be printed on one side only and numbered, and a header or footer will identify the offeror's name and the RFP number. Each proposal volume copy will be submitted in a three-ring binder. As desired, attachments can be provided to amplify proposal Volume I Sections 2 through 5. However, the attachment pages will be counted against the 40-page limitation.

3.0 EVALUATION CRITERIA

The Volume I proposal will be evaluated by the contract administrator using the following major evaluation topics:

Evaluation Topic 1. *Did the proposal provide sufficient details?*
Evaluation Topic 2. *Was the proposal readable, credible, and persuasive?*
Evaluation Topic 3. *Did the proposal properly respond to the RFP requirements?*
Evaluation Topic 4. *Did the proposal properly comply with the proposal instructions in the RFP?*

Evaluation details are provided in Attachment (1) of this RFP.

RFP Attachment (1)

Proposal Evaluation Criteria
(Note: The following provides evaluation guidelines for instructor and student use. Read and evaluate each proposal. Grade each subtopic using a 1 to 5 grading scale: 1 = poor, 2 = fair, 3 = good, 4 = very good, 5 = excellent.)

Major Evaluation Topic 1. *Did the proposal provide sufficient details?*

Evaluation Subtopics (1–5 grade):

1.1 Who will do the work?
1.2 What work will be done?
1.3 When will the work be done?
1.4 Where will the work be done?
1.5 Why is the work necessary?
1.6 How will the work be done?

Average grade of evaluation subtopics:
Major evaluation topic strengths:
Major evaluation topic weaknesses:
Major evaluation topic recommendations:

Major Evaluation Topic 2. *Was the proposed readable, credible, and persuasive?*

Evaluation Subtopics (1–5 grade):

2.1 Was the writing clear, understandable, and logical? Did it use proper grammar? Was there a consistent writing style?
2.2 Were the proposed approaches credible and effectively substantiated?
2.3 Were the proposed approaches clearly linked to specific benefits for the customer?
2.4 Was the proposal visually appealing? For example, was the layout effective; were graphics effectively located; and were main points highlighted with devices such as boxes and bulleted listings?
2.5 Did the proposal effectively use graphics for simplifying and reinforcing ideas and for stimulating interest?
2.6 Did the text effectively refer to and discuss the graphics?

Average grade of evaluation subtopics:
Major evaluation topic strengths:
Major evaluation topic weaknesses:
Major evaluation topic recommendations:

Major Evaluation Topic 3. *Did the proposal properly respond to the RFP requirements?*

The proposal will also be evaluated for (1) the *soundness* of the *approach* it offered to meet RFP requirements (a qualitative grade), (2) its *understanding* of the *RFP requirements* as reflected in its content and offered approach (qualitative grade), (3) its *coverage* of *all* pertinent *RFP requirements* (quantitative grade, not qualitative), and specifically (4) the *support* of its offered approach in meeting the

PAS schedule requirements (qualitative grade). The following sections of the proposal will undergo this evaluation:

- Task Approach
- Problem Assessment
- Study Management
- Personnel Résumés

3.1 Task Approach

Evaluation subtopics (1–5 grade)

3.1.1 Soundness of the approach
3.1.2 Understanding of the RFP requirements
3.1.3 Coverage of all RFP requirements
3.1.4 Support of PAS schedule requirements

Average evaluation subtopic grade:
Task approach strengths:
Task approach weaknesses:
Task approach recommendations:

3.2 Problem Assessment

Evaluation subtopics (1–5 grade)

3.2.1 Soundness of the approach
3.2.2 Understanding of the RFP requirements
3.2.3 Coverage of all RFP requirements
3.2.4 Support of PAS schedule requirements

Average evaluation subtopic grade:
Problem assessment strengths:
Problem assessment weaknesses:
Problem assessment recommendations:

3.3 Study Management

Evaluation subtopics (1–5 grade)

3.3.1 Soundness of the approach
3.3.2 Understanding of the RFP requirements
3.3.3 Coverage of all RFP requirements
3.3.4 Support of PAS schedule requirements

Average evaluation subtopic grade:
Study management strengths:
Study management weaknesses:
Study management recommendations:

3.4 Personnel Résumés

Evaluation subtopics (1–5 grade)

3.4.1 Soundness of the approach
3.4.2 Understanding of the RFP requirements
3.4.3 Coverage of all RFP requirements
3.4.4 Support of PAS schedule requirements

Average evaluation subtopic grade:
Personnel résumé strengths:
Personnel résumé weaknesses:
Personnel résumé recommendations:

Major Evaluation Topic 4. ***Did the proposal properly comply with the proposal instructions in the RFP?***

Answer with a yes or no; if no, indicate the deficiency.

Index

Note: Page entries in *italics* refer to figures; entries followed by *t* refer to tables.